AF888405

HARVESTED MINDS

The Captive Golden Goose

Author: Jordan A. Cross

First Edition

ISBN: 978-1-967440-11-5

AIMQWEST BOOKS®
AIMQWESTBOOKS.COM

Foreword by the author

There are moments in history when the silence of complacency proves more dangerous than the chaos of dissent. We are living in such a moment. This book was not conceived as an abstract exercise in theory, nor as a detached chronicle of the digital age. It was born from the lived experiences of countless innovators, artists, immigrants, and visionaries who dared to believe that their creativity and labor would be honored on their own terms—only to discover, often too late, that they had been quietly captured by a system that thrives on dependency.

The fable of the golden goose has endured for centuries as a parable of greed. Yet in our age, the metaphor has taken on a darker inversion. Today, the goose is not slaughtered. It is contained—fed, sheltered, and celebrated, but never free. The platforms that dominate our digital and economic lives have perfected the art of extracting perpetual value while ensuring that the producers of that value remain tethered, never permitted to own their labor fully nor to walk away without devastating loss.

In these pages, you will encounter stories that are not hypothetical. They are composites drawn from real trajectories, lived by those whose names seldom appear in headlines yet whose work sustains industries and empires. These individuals are the lifeblood of the digital economy, and yet their autonomy has been bartered away through contracts disguised as opportunities, algorithms cloaked in neutrality, and promises framed in the language of empowerment.

This book is not merely an exposé. It is a map—of the cage, of its mechanics, and of the paths that still remain open to those who would reclaim their sovereignty. It is intended for the coder whose innovations are siphoned into proprietary ecosystems, the musician whose art is reduced to fragments for algorithmic playlists, the immigrant entrepreneur whose dreams are repackaged as "features" in someone else's quarterly earnings report. It is also for the reader who senses that something is amiss in the relationship between creativity and capital, but who has lacked the language to articulate it.

If you recognize yourself in these pages, know this: you are not alone. The architecture of captivity relies on isolation, on making each creator believe their struggle is unique, their dependency inevitable. It is not. The same systems that confine can be studied, resisted, and dismantled. The purpose of this work is not to romanticize rebellion, but to illuminate strategies—legal, technical, financial, and communal—that restore power to its rightful origin: the creator.

The age of the captive golden goose is nearing its twilight. Whether its end comes through conscious liberation or through the obsolescence imposed by artificial intelligence is not yet decided. What is certain, however, is that history will not look kindly on those who surrendered their sovereignty without a fight. My hope is that these chapters offer both clarity and courage—clarity to see the invisible chains for what they are, and courage to step beyond them.

The story of captivity has been written for too long. The story of sovereignty begins here.

Jordan A. Cross

TABLE OF CONTENTS

Introduction,
"Recognizing the Cage"

Opening vignette

Under the soft haze of a streetlamp in a crowded immigrant neighborhood, Lucien worked late into the night, his small apartment doubling as both home and laboratory. A first-generation immigrant from France, he had arrived in the United States with little more than a degree in computer science, a head full of ideas, and an unshakable belief that the internet was the great equalizer. By day, he worked a modest job as a systems analyst for a regional logistics company; by night, he poured his energy into developing an AI-driven language learning platform that could adapt to the cultural and dialectal nuances of its users. The application, unlike the sterile and one-size-fits-all models dominating the market, learned from the learner, shifting its responses in real time to honor idioms, regional sayings, and even the rhythm of the user's speech.

The breakthrough came when a demo clip of his software went viral on a professional networking site. In a matter of days, Lucien was fielding messages from venture scouts, platform representatives, and "innovation program coordinators" from major tech companies. The language of the invitations was disarmingly generous: mentorship, funding, exposure, technical support. Each offer came wrapped in the rhetoric of empowerment. Yet beneath the carefully curated kindness, there was an unspoken urgency, get in now, integrate now, before someone else does.

He chose to partner with one of the largest platforms in the world, drawn in by the promise of access to their colossal user base. The onboarding process was a whirlwind of nondisclosure agreements, intellectual property clauses, and "standard" monetization terms. He was assured these were formalities, mere protective measures to keep bad actors at bay. The reality was more insidious. Within months, the platform's algorithms began subtly reshaping his software's interface, dictating UI changes under the guise of "user experience optimization." His code was never taken outright, but new, remarkably similar features began appearing in the platform's native tools, bundled for free to premium subscribers.

Revenue trickled in from the platform's monetization model, but the payout schedule shifted unpredictably, sometimes delayed by weeks without explanation. Marketing visibility, once promised as part of the deal, was curtailed unless he participated in "sponsored boosts" that ate away at his slim margins. When he attempted to roll out an independent web version of his product to build a direct customer base, his traffic from the platform's referral channels dropped precipitously. At first, he thought it was coincidence, but a sympathetic contact inside the company confirmed what he had suspected: independent distribution triggered a downgrade in the platform's visibility algorithms.

Lucien was not alone. In private online forums for minority tech entrepreneurs, similar stories were whispered, each with its own variations, but all sharing the same anatomy. A promising innovation gains traction. The creator is courted by a major platform with promises of partnership. The product is integrated, reshaped, and, piece by piece, absorbed. The creator's control narrows, the monetization becomes erratic, and any attempt to escape is met with invisible penalties. The higher the potential market impact, the tighter the leash.

For Lucien, the most painful realization was that his work had indeed reached millions of users worldwide, but not on his terms. The branding had shifted to the platform's aesthetic, the data was routed through their proprietary analytics, and the cultural sensitivity modules he had painstakingly built were stripped down into market-friendly "regional packs" sold under the platform's umbrella. He received no royalties for these additions; legally, they were considered "derivative features" developed in-house. His unique vision, born from lived experience, from the memory of his grandmother teaching him French proverbs under the warm light of a rustic kitchen, had been reduced to a line item in a multinational's quarterly growth report.

The weight of it was not just financial. There was a quiet psychological toll to knowing that his creation was shaping language learning for an entire generation, yet his name was absent from the credits. Invitations to speak at conferences came, but only under the condition that he appear as a "platform innovator" rather than an independent founder. His own narrative was being rewritten in real time, curated to reinforce the myth of corporate benevolence rather than the truth of creative captivity.

As the years passed, Lucien's nights grew quieter. He still coded, but the fire that once fueled his work had dimmed. The dream had not died, it had been captured, commodified, and repackaged until it no longer belonged to him. Yet deep down, he knew the story was not over.

Others like him were beginning to see the contours of the same trap, to recognize that what looked like partnership was, in fact, a velvet cage. And in that recognition lay the first spark of something dangerous to those who profited from it: the will to break free, to take back what was theirs, and to ensure that the next golden goose would not lay its eggs in captivity.

It was during one of his rare visits back to his hometown that the spark reignited. Sitting in a crowded café, Lucien overheard a group of young students huddled over a battered laptop, laughing as they tried out an online language-learning tool. He recognized fragments of his own interface, small design decisions no one else would have made, woven into the platform's software. They were excited, engaged, and unaware that the soul of what they were using had come from the very streets around them.

The scene was bittersweet. He saw in their faces the same hunger for opportunity that had driven him, the same belief that knowledge could unlock the world. But he also understood that unless something changed, these students could one day find themselves on the other side of the same bargain, creators of value they did not own, innovators whose names and legacies would be eclipsed by corporate branding.

Back in the United States, Lucien began quietly rebuilding. He reached out to other immigrant and minority innovators who had endured similar fates. Together, they began mapping the patterns, how "exclusive beta programs" subtly replaced creator autonomy with dependency; how technical integrations became permanent entanglements; how clauses buried deep in standard contracts gave platforms sweeping rights over "derivative innovations." The more they shared, the clearer it became: these were not isolated cases, but a repeatable, scalable system of capture.

It was no longer just about reclaiming his own work, it was about dismantling the mechanisms that allowed such exploitation to flourish. Lucien realized that the only way forward was to build beyond the reach of the gatekeepers. He began experimenting with decentralized hosting, blockchain-based licensing for intellectual property, and peer-to-peer distribution channels. The goal was not to reject technology, but to reimagine it in a way that honored the sovereignty of its creators.

The process was slow, and the resistance was real. Platforms attempted to undercut his relaunch by pushing free alternatives that mimicked his new features. Some contacts quietly warned him that continuing down this path could result in being blacklisted from certain industry networks. But the more the pushback intensified, the more he knew he was hitting a nerve.

And then, unexpectedly, came validation. A coalition of independent educators in South Asia, frustrated with the limitations and costs of major learning platforms, adopted his tool in its decentralized form. This time, there were no opaque algorithms throttling access, no sudden changes in monetization, no branding sleight of hand. His name was there, his vision intact, and his income direct. Word spread, not virally in the shallow sense of the term, but steadily, through trusted community channels that did not require corporate mediation.

The experience reminded him of a truth he had nearly forgotten: freedom was rarely granted; it had to be taken. He saw clearly now that the platforms' power did not just come from their technology or their capital, but from their ability to isolate creators, to make them believe they were alone in their struggles. Breaking that isolation, building alliances, sharing information, refusing to accept the inevitability of captivity, was the first step toward dismantling the system itself.

In the end, Lucien's journey was not a fairy tale of perfect justice. The platform that had once courted and contained him continued to thrive, profiting from his early innovations. But he had carved out something they could not touch: an independent ecosystem where his work lived on its own terms, reaching the people it was meant to serve without passing through a corporate filter. And in the quiet moments, when he saw a message from a student halfway around the world who had learned a new language through his tool, he felt the weight of the cage lift just a little more.

He understood that the fight was far from over. But he also knew that every creator who refused to become another captive golden goose brought the end of the era closer. Somewhere in another city, another small apartment, another late-night coder was building the next great thing. And perhaps this time, when the knock came from the velvet cage, they would know to leave the door unopened.

Setting the frame

The image of the golden goose, drawn from the timeless fable, speaks to the relationship between a rare and irreplaceable source of value and those who control it. In the original tale, the goose lays eggs of solid gold, a miracle of nature that could sustain its keeper indefinitely. The moral, often interpreted as a warning against greed, hinges on the folly of killing the goose to seize all its wealth at once, destroying the source of abundance in the pursuit of immediate gain. In the modern digital economy, this metaphor has taken on a subtler and more complex meaning. High-value producers of intellectual property, whether artists, developers, engineers, researchers, or content creators, are the living geese in this analogy, and their "golden eggs" are the innovations, designs, and cultural artifacts they produce. These outputs carry immense and recurring economic potential, but unlike the static fable, the eggs here are infinitely reproducible, scalable, and capable of generating continuous profit for as long as they can be extracted and controlled.

Yet in this reimagined parable, the golden goose is seldom killed outright. The exploitation lies not in a single fatal act, but in the careful management of the goose's environment to ensure that it continues to produce on terms that benefit the keeper. Instead of a barnyard fence, the modern captor builds an ecosystem of dependencies, platforms, tools, distribution networks, and monetization schemes, that make the producer's continued output inseparable from the captor's infrastructure. The goose is kept alive and productive, but its freedom to roam, to sell its own eggs on its own terms, is quietly removed. Over time, the goose may even forget that it once had such freedom, believing the constraints to be the natural order of things.

For high-value intellectual property producers, the metaphor extends into the deliberate construction of systems that both cultivate and confine. A digital artist may be granted premium exposure in an online marketplace but only in exchange for exclusive hosting rights, thereby surrendering the ability to distribute independently. A software developer may be given access to cloud computing credits or marketing boosts, only to find that every line of code created in that environment is subject to a platform's license, whether or not the developer intended it. The apparent generosity mirrors the care of the fable's farmer, feeding, sheltering, and tending to the goose, while concealing the primary objective: sustained extraction of value.

The metaphor also reveals why the most gifted and original producers are often the most aggressively targeted. Just as a farmer would never lavish resources on an ordinary goose when a golden one exists in the coop, platforms and corporate entities concentrate their attention on those with a proven record of generating breakthroughs. Algorithms identify rising stars in creative or technical fields, predictive analytics forecast their future productivity, and outreach teams move swiftly to secure them before competitors can. This early capture ensures that the "eggs" will mature within a controlled environment, preventing the possibility that their value will flow to independent markets or alternative distribution channels.

What distinguishes the digital golden goose from its fabled ancestor is the scale and subtlety of the apparatus. In the globalized knowledge economy, a single producer's output can be replicated to billions of devices, embedded in countless derivative works, and monetized across multiple revenue streams, advertising, subscriptions, licensing, data collection, without the producer ever holding more than a fractional share of the returns. Here, the eggs are not physical treasures stored in a box, but lines of code, original scores, patented designs, viral media, or proprietary datasets, all seamlessly integrated into a larger commercial machine. The producer may receive periodic payments or token recognition, but the vast majority of the economic and strategic benefit remains with the platform.

In this context, the moral of the original tale is inverted. In the fable, killing the goose ends the wealth; in the digital age, the captor's aim is to keep the goose alive indefinitely, producing without interruption, but never gaining the capacity to own or fully profit from the eggs. This inversion is why the metaphor resonates so powerfully with high-value IP producers: it captures both the dependency and the extraction, the sense of being indispensable yet dispossessed.

The enduring power of the golden goose story lies in its capacity to make visible what is otherwise hidden in the smooth language of corporate partnership and platform support. The gilded cage is not presented as a prison, but as a palace. The feed is plentiful, the shelter is comfortable, and the pathways out are obscured or made to appear perilous. But the fundamental truth remains: as long as the goose cannot take its eggs elsewhere, it is not free. Recognizing this dynamic is the first step toward dismantling it, for no amount of care or reward can compensate for the loss of sovereignty over one's own creation.

This understanding reframes the conversation around creator-platform relationships. It is not merely about whether a producer is paid fairly for a given work, but about whether they retain the structural capacity to determine how, where, and under what terms their intellectual property is used. The "golden goose" metaphor clarifies that value extraction is not confined to the moment of sale; it is embedded in the very architecture of ownership, licensing, and distribution. Each term of service, each technical dependency, each market gatekeeping function is a strand in a larger net designed to keep the goose laying for the benefit of the captor.

Consider the case of algorithmic visibility. In the fable, the farmer decides when and how often the goose lays its eggs, perhaps through feeding or breeding control. In the digital realm, platforms manipulate discoverability, engagement metrics, and monetization thresholds. The creator may believe that their visibility rises and falls with audience interest, but in truth, these shifts are engineered to incentivize behaviors beneficial to the platform's revenue model. Increased output, greater exclusivity, and deeper integration into proprietary tools all serve to keep the eggs flowing steadily into the platform's collection, ensuring that the producer remains reliant on the same systems that limit their autonomy.

This is why the metaphor also functions as a diagnostic tool for identifying exploitation. If a creator's work is disproportionately enriching others while leaving them unable to achieve meaningful independence, they are likely inside the gilded cage of a modern goosekeeper. The test is simple yet profound: could the creator leave the platform tomorrow, take their audience or customers with them, and sustain themselves without the platform's infrastructure? If the answer is no, the goose is captive, no matter how prestigious the environment or generous the perks.

Moreover, the metaphor underscores the moral hazard for those who benefit from the goose's captivity. In the fable, greed blinds the farmer to the long-term sustainability of their wealth. In the digital age, the captor's greed manifests in the aggressive consolidation of market power, the suppression of alternative channels, and the systematic rewriting of ownership norms to favor intermediaries over originators. Left unchecked, this dynamic does more than harm individual producers; it erodes the diversity and resilience of the creative and technological ecosystems themselves. When only platform-approved eggs can reach the market, the world is deprived of innovations that might have flourished outside the cage.

Finally, the golden goose analogy holds a warning for the producers themselves. It is easy to mistake care for captivity when the captor's hand also feeds. Comfort, predictability, and steady, if limited, rewards can dull the instinct to seek freedom. Many creators remain in gilded cages not because they are unaware of their constraints, but because the risk of leaving seems greater than the cost of staying. Yet, as the fable reminds us, the goose's true value lies not in the eggs it produces for others, but in its inherent capacity to create. Until that capacity is reclaimed and directed toward the producer's own ends, the promise of independence remains an illusion, and the story remains incomplete.

The invisible capture

The most sophisticated forms of exploitation rarely arrive in the guise of coercion. They appear instead as offers of advancement, carefully tailored packages of resources, visibility, and prestige that speak directly to a creator's ambitions. This is the essence of invisible capture: a process by which control is transferred without the subject fully realizing it, through agreements framed as opportunities and relationships branded as partnerships. The outward trappings are benign, even flattering. Invitations arrive with promises of amplification, technical support, and access to audiences that once seemed unreachable. The language is saturated with the rhetoric of empowerment, inclusion, and innovation. The intended effect is to blur the line between help and possession until the distinction becomes meaningless.

In such scenarios, the creator's guard is naturally lowered. A new distribution agreement might come wrapped in the prestige of association with a household name. A premium account or beta program might be pitched as a reward for demonstrated talent. The contract terms are positioned as standard industry practice, the exclusivity clause as a badge of honor. Every element is designed to frame the transaction as mutually beneficial, obscuring the asymmetry in bargaining power and the long-term consequences of surrendering certain rights. Over time, these "benefits" begin to alter the creator's operational reality, where they publish, how they monetize, and to whom they owe their primary allegiance, until their capacity to function independently has been quietly eroded.

What makes invisible capture so insidious is that it often exploits genuine short-term gains to mask long-term losses. A musician offered an advance on royalties may be able to finance their next project immediately, but the licensing structure could ensure that the bulk of revenue from future streams accrues to the platform indefinitely. A developer given free access to proprietary APIs can scale their application rapidly, yet find themselves unable to migrate to open infrastructure without dismantling their product entirely. By the time the creator recognizes the cost, the dependencies are too deeply embedded to unwind without sacrificing livelihood, audience, or both.

The illusion of opportunity is reinforced by the cultural narrative that visibility equals success. Platforms understand that recognition carries its own intoxicating power. Being featured on the front page of an app store or showcased in a curated campaign creates a sense of validation that can outweigh concerns about autonomy. This psychological reinforcement serves as both carrot and conditioning: the creator internalizes the belief that platform approval is not only desirable but essential. In doing so, they may willingly adjust their creative choices to align with the tastes and policies of the gatekeeper, mistaking compliance for collaboration.

Invisible capture also thrives on the strategic use of opacity. Contractual language is dense and ambiguous, with key operational details buried in appendices or deferred to "policy updates" that can be amended unilaterally. Technical systems are presented as neutral and automated, even when they are calibrated to favor content and products that generate higher returns for the platform, regardless of their intrinsic merit. The creator, deprived of full visibility into the mechanisms governing their reach and revenue, must rely on the explanations provided by the very entity whose interests may be in conflict with their own.

There is a deeper cultural dimension to this phenomenon: in a globalized, hyper-competitive economy, creators are conditioned to see any foothold as worth taking, any alliance as worth securing. The scarcity mindset, fueled by the reality that only a fraction of high-value producers achieve substantial market penetration, makes it easier for platforms to frame restrictive agreements as rare chances not to be missed. The urgency of "limited spots" or "exclusive cohorts" suppresses due diligence, replacing caution with gratitude. When exploitation is framed as a gift, rejecting it can feel not only risky but ungrateful.

The endpoint of invisible capture is not open domination but quiet assimilation. The creator's brand, process, and product remain visible, but they exist entirely within the architecture of the captor. Audiences encounter the work as a feature of the platform rather than the independent creation of its originator. Over time, the platform becomes the default point of reference, and the creator's identity is subsumed into a broader corporate narrative. By then, the transition from autonomy to captivity has been so gradual, and so well-disguised as progress, that many accept it as the natural trajectory of success.

Why this book exists

This work exists because the patterns it exposes are too often rendered invisible by the very systems that depend on their concealment. In an age when creative and intellectual labor can be replicated at infinite scale, the mechanisms of control have evolved beyond overt theft or brute-force appropriation. They now operate through finely tuned frameworks of dependency, contractual sleight of hand, and psychological conditioning that recast exploitation as collaboration. The purpose of this book is to dismantle those illusions, to equip high-value intellectual property producers with the clarity and tools needed to recognize the architecture of their own captivity, and to map out viable paths toward independence without romanticizing the risks involved.

Its scope extends beyond a single industry or creative discipline. Whether the reader is a developer, musician, researcher, writer, or designer, the dynamics at play are remarkably consistent. Algorithms and corporate outreach teams identify promising talent early, offering access to markets, tools, and audiences that would be impossible to reach alone. In exchange, the producer yields not only a percentage of their output, but, more consequentially, the capacity to determine its distribution, monetization, and evolution. This exchange is rarely framed as a loss of sovereignty; instead, it is couched in the aspirational language of growth and partnership, ensuring that even those who suspect a hidden cost feel compelled to accept the bargain.

The reader will gain, first and foremost, a vocabulary for naming the dynamics they may already be experiencing but have struggled to articulate. Understanding concepts such as "invisible capture," "dependency loops," and "derivative control" transforms vague unease into concrete awareness. That awareness is essential because the structures of exploitation thrive in ambiguity. Once the dynamics are visible, the reader can begin to evaluate their position with precision, identifying the points at which the perceived benefits of a platform relationship begin to erode their long-term autonomy.

Beyond diagnosis, this book offers a framework for action. It does not suggest that disengagement from major platforms is always feasible or strategically sound; rather, it focuses on the deliberate cultivation of parallel channels of control. This may mean diversifying distribution networks, securing legal protections over intellectual property, or leveraging decentralized technologies to mitigate the risks of single-platform dependency. The goal is not a wholesale rejection of digital infrastructure but the construction of a position from which no single actor can dictate the terms of a creator's survival.

By reading, the audience will also gain an understanding of the systemic forces that make certain groups, particularly minorities, immigrants, and those without institutional backing, more vulnerable to these forms of capture. This awareness is not merely academic; it informs practical strategies for resilience, community-building, and resource-sharing that can shift the balance of power. Recognizing that individual independence is often inseparable from collective empowerment, the book situates personal liberation within a broader movement toward structural reform.

Ultimately, the value of this work lies in its commitment to revealing the gap between the promise of digital opportunity and its lived reality for those whose talent and innovation are the raw materials of the modern economy. It is written not to persuade creators to reject every offer of collaboration, but to ensure they enter each negotiation with their eyes fully open, armed with the knowledge that the true measure of a partnership is not the prestige of the association but the degree to which it preserves the creator's control over the fruits of their own labor.

Overview of structure

This work unfolds in a deliberate sequence, guiding the reader from initial recognition of the problem to the development of concrete strategies for liberation. It begins by illuminating the hidden architecture of exploitation, those systemic forces that quietly bind high-value intellectual property producers to the very entities that profit most from their work. The first movement dissects the anatomy of such systems, tracing how large technology platforms, corporate intermediaries, and institutional networks identify promising creators, draw them into seemingly beneficial arrangements, and transform those arrangements into enduring dependencies. Through this analysis, the reader will encounter the psychological and technical levers, ranging from targeted outreach to algorithmic manipulation, that make disengagement not only difficult but often economically perilous.

Once this anatomy is laid bare, the focus shifts to the process of "marking" the most valuable creators. Here, the discussion examines the early warning signs that a producer has entered a platform's high-value watchlist: invitations to exclusive programs, subtle restrictions on independent monetization, and the harvesting of behavioral and creative data to anticipate and influence future output. This part also addresses the structural inequities that render certain demographics, particularly immigrants, minorities, and those without institutional backing, more vulnerable to capture, contextualizing the personal within the systemic.

The exploration then turns to the mechanisms of control, described as the tools of enslavement. These include contractual vagueness that allows for arbitrary policy enforcement, technical means of limiting reach or functionality, financial throttles that undermine stability, and psychological tactics designed to erode a creator's confidence in their ability to succeed independently. A detailed case study traces one subject's journey through multiple platforms, offering a longitudinal view of how these tools operate in concert over time.

From there, the narrative addresses the economic dimension of captivity. This includes quantifying the gap between the value a creator generates and the share they actually receive, exposing the closed-loop systems that keep both production and distribution within platform-controlled channels, and calculating the opportunity costs of blocked or delayed ventures. It also considers how captive intellectual property distorts broader markets, reshaping norms in ways that disadvantage independent actors.

In the following phase, the book dismantles the carefully crafted narratives that sustain these exploitative structures. It analyzes the rhetoric of "empowerment" and "partnership" that masks unequal power dynamics, exposes the decision-makers and automated systems that determine creator outcomes, and presents stories of those who have successfully escaped the confines of platform control.

With the problem fully mapped, the final third of the book is devoted to solutions. This includes practical frameworks for auditing one's dependencies, building parallel infrastructures that enable self-reliance, negotiating stronger legal protections, and securing funding without surrendering control. The strategies emphasize not only technical and legal measures but also the cultivation of community alliances and the use of decentralized technologies to reduce vulnerability to gatekeeping.

The journey concludes with a forward-looking vision: what a post-captive creative economy could look like, the steps required to bring it into being, and a rallying call for sustained vigilance. By the end, the reader will have traversed a complete arc, from recognizing the cage to holding the keys, and will be equipped with both the insight and the tools to ensure their creative sovereignty is not only reclaimed but preserved.

Chapter 1, The Anatomy of a Captive System

Identifying the Players

The network of actors that sustains the captivity of high-value intellectual property producers is neither accidental nor loosely organized. It is a structured, interdependent web of entities whose roles and incentives align toward a singular objective: the continuous inflow of fresh intellectual capital that can be monetized at scale. At the forefront are the global technology platforms, massive digital ecosystems whose names dominate every conversation about reach, engagement, and discoverability. These entities are not merely service providers; they are vertically integrated empires controlling infrastructure, distribution, analytics, and monetization in ways that make their dominance self-reinforcing. They operate social networks, app stores, content marketplaces, cloud services, and advertising exchanges under one corporate roof, creating a closed-loop environment where the creator's journey from idea to audience remains entirely within their jurisdiction.

Their business models are designed to absorb a relentless stream of new intellectual property. Without this influx, their ability to attract users, advertisers, and investors would erode. User engagement, the metric upon which their stock valuations often hinge, is fueled by novelty, new content, new applications, new cultural touchpoints. Creators are the primary suppliers of this novelty, and the platforms' algorithms are tuned to maximize both the extraction of value from each contribution and the dependency of the contributor. They do so through a combination of algorithmic curation, behavioral nudging, and contractual frameworks that encourage continuous production while discouraging migration to alternative venues. In this way, the flow of IP becomes as critical to the platform's survival as the flow of raw materials is to a manufacturing plant.

Beneath the visible surface of these dominant players lies a second tier: the corporate ecosystems that orbit and integrate with them. This includes major entertainment conglomerates, e-commerce giants, financial services firms, and enterprise software providers whose growth strategies increasingly hinge on embedding themselves into the core functions of the tech behemoths. They build tools, services, and APIs that lock creators into particular workflows, ensuring that once an artist, developer, or educator begins operating within their environment, the switching costs become prohibitively high. These corporate allies may appear to offer independence, white-label storefronts, licensing agencies, content syndicators, but their revenue streams are often inextricably tied to the very platforms they claim to help navigate. In effect, they extend the reach of the primary captors, multiplying the points of control while maintaining the facade of diversity in the marketplace.

Adding yet another layer to this structure are the institutional intermediaries: venture capital firms, talent agencies, academic institutions, and nonprofit accelerators that ostensibly exist to nurture and champion creative and technical talent. While many operate with genuine intent to empower, their structural incentives frequently align with the consolidation of creator output into the hands of larger, better-capitalized entities. For example, a venture-backed startup may provide funding and infrastructure to a promising developer, but its exit strategy, whether acquisition by a major platform or integration into its ecosystem, often results in the founder's IP becoming a corporate asset. Similarly, talent agencies may negotiate distribution deals that maximize short-term exposure but commit artists to multi-year exclusivity agreements with opaque royalty structures. Even universities and research institutions, under pressure to demonstrate commercial impact, can become conduits funneling breakthrough innovations into private hands long before the inventor has had the opportunity to negotiate equitable terms.

Each of these actors, from the platform giants to the seemingly peripheral intermediaries, depends on a constant pipeline of intellectual property to sustain their competitive position. The nature of IP as an asset class makes this dependency particularly acute: unlike physical goods, it can be reproduced at negligible marginal cost, allowing for near-limitless scalability once captured. But its value decays rapidly without replenishment, as audiences and markets continually demand the next new thing. This creates an insatiable hunger for talent and output, driving these entities to develop increasingly sophisticated methods for identifying promising creators early, integrating them into their ecosystems, and structuring the relationship so that the flow of IP remains uninterrupted.

In practice, this means that the players in this system are not competing primarily for consumers, they are competing for producers. The public narrative may focus on user acquisition and market share, but internally, the most strategic conversations revolve around creator acquisition and retention. The most successful captors are those who have mastered the dual art of attraction and containment: attracting high-value producers with offers of resources, exposure, and technical support, while containing them through interlocking dependencies that make departure economically irrational or technically infeasible. The result is a market that appears vibrant and diverse on the surface, yet beneath it is structured around a handful of central nodes that quietly control the flow of the most valuable commodity in the digital age: the ingenuity of human creators.

What makes this network especially resilient is the way its members reinforce each other's positions while appearing to compete. The largest platforms, social media giants, search engines, app stores, project the image of fierce rivalry, yet they share remarkably similar structural goals: to centralize user activity, capture creator output, and channel monetization through their own systems. Where direct partnerships are possible, they create formal alliances, such as cross-platform authentication or co-branded initiatives, which extend the reach of each participant's data collection and advertising apparatus. Where formal collaboration is not viable, they engage in tacit coordination, mimicking each other's features and business models so that a creator leaving one ecosystem encounters the same constraints in another. The effect is a form of competitive homogenization: different brand names, identical cages.

The corporate ecosystems that orbit these giants act as both feeders and filters. Many of them specialize in scouting and cultivating talent that shows early signs of high-value potential. They may present themselves as enablers of independence, but their commercial survival depends on channeling the most promising producers toward larger players that can monetize them at scale. In the music industry, for example, independent labels often function less as permanent homes for artists and more as talent farms for major record companies, whose digital distribution arms are themselves deeply integrated into the largest streaming platforms. In the software world, niche developer tools and frameworks may be acquired by a tech conglomerate within months of gaining traction, instantly binding their existing user base to the parent company's infrastructure. Even in publishing, small imprints are often owned by multinational conglomerates whose digital distribution deals tie them to dominant e-commerce marketplaces.

Institutional intermediaries serve a subtler but equally important role in sustaining the cycle. Venture capital firms, under pressure to deliver high returns, steer startups toward exit paths that prioritize liquidity events, acquisitions and mergers, over long-term independence. The startups themselves often become vehicles for harvesting innovative IP, whether in the form of proprietary algorithms, platform-specific content formats, or novel distribution methods. Talent agencies, operating at the intersection of entertainment, sports, and digital media, use their negotiation leverage to secure broad rights for their clients but frequently trade those rights in bulk to media conglomerates in exchange for immediate visibility and promotional budgets. Academic and research institutions, especially those reliant on corporate sponsorships, may place intellectual property into exclusive licensing agreements that effectively transfer the commercial benefits of publicly funded innovations to private actors with minimal oversight.

Underpinning all of these relationships is the fact that intellectual property, unlike raw materials or manufactured goods, gains much of its value through the networks and infrastructures that distribute it. The players in this system control those infrastructures, from content delivery networks to recommendation algorithms, from payment gateways to analytics dashboards. By shaping the conditions under which IP reaches audiences, they can dictate not only who profits but also which kinds of creativity and innovation are deemed viable in the marketplace. This control allows them to manage risk at the macro level, promoting content that aligns with their commercial objectives while quietly suppressing or deprioritizing work that might disrupt their revenue models or challenge their dominance.

The continuous influx of IP that fuels these entities is therefore not a passive occurrence; it is the product of active cultivation. Scouting algorithms sift through oceans of data to identify emerging trends, breakout personalities, and high-engagement formats long before they appear on the radar of the general public. Business development teams move quickly to offer partnerships, grants, or feature placements that appear to be recognition of merit but are in fact strategic investments in securing exclusive or semi-exclusive access to the creator's pipeline. Legal and policy teams design terms of service and licensing agreements that anticipate future monetization opportunities, often granting the platform perpetual or transferable rights over any derivative works created within its environment.

What is perhaps most striking is how normalized this entire process has become. For a new creator, entering the orbit of one of these major players is often framed as the ultimate goal, a milestone that validates their skill and signals their arrival in the industry. The infrastructure of dependence is disguised as a badge of legitimacy, and the consolidation of IP under a handful of powerful actors is marketed as efficiency, scalability, or global reach. Few stop to question the systemic fragility this creates: when the same small set of entities control distribution channels, payment systems, and audience access, the creative economy becomes vulnerable to policy changes, corporate mergers, or shifts in investor sentiment that can instantly reshape the livelihood of millions.

These players, diverse in their public identities but unified in their structural aims, have mastered the art of presenting themselves as indispensable partners while functioning, in effect, as gatekeepers and landlords in the digital commons. Their survival depends on ensuring that creators continue to see them as both opportunity and necessity, a perception reinforced by an ecosystem of dependent corporations and intermediaries whose fortunes rise and fall with the flow of intellectual property into the system. This is the architecture of modern capture: a vast, interlocking network in which every path forward for a creator seems to lead through the same guarded gates, and where the act of creation itself becomes a resource continuously extracted to feed the machine.

The durability of this architecture is further reinforced by the asymmetry of resources between the captors and the captured. The major platforms and their corporate allies have at their disposal vast legal teams, sophisticated data analytics, and financial reserves that allow them to operate on strategic timelines far longer than those of individual creators. They can afford to cultivate relationships for years before fully absorbing a creator's output, secure in the knowledge that the creator's economic realities will often compel them to accept terms they might otherwise reject. This patience is strategic; it ensures that the moment of capture feels like the culmination of a mutually beneficial courtship rather than the conclusion of a calculated acquisition.

At the heart of this system lies a fundamental economic truth: intellectual property in the digital era functions as both currency and collateral. Every new song, application, film, dataset, or design not only has direct market value but also serves as a bargaining chip in the competition for user engagement, advertiser spending, and investor confidence. The companies that dominate this landscape have structured themselves to extract maximum value from both aspects. On the one hand, they monetize IP directly through sales, subscriptions, and licensing. On the other, they leverage it indirectly to enhance their own market position, by increasing platform stickiness, attracting complementary developers or partners, and reinforcing their status as indispensable hubs for cultural and technological life.

This dual monetization model is why the continuous influx of IP is existential to their business. Without a steady stream of new work, the appeal of their platforms diminishes, user engagement wanes, and advertisers redirect their budgets elsewhere. For investors, a slowdown in content or innovation is interpreted as a warning sign, leading to reduced valuations and diminished leverage in mergers, acquisitions, and strategic partnerships. As a result, the incentive to secure IP at the earliest possible stage is built into the very DNA of these organizations. It is not enough for them to acquire finished works; they seek access to the creative process itself, integrating their tools and services into the workflows of creators so that the raw material of innovation flows directly into their systems long before it reaches the market.

The corporate ecosystems and institutional intermediaries that support this model act as amplifiers of both reach and control. Their function is not merely to bring creators into contact with the platforms, but to align the creator's goals with the platform's commercial imperatives. This alignment is achieved through carefully framed narratives about scalability, market access, and professionalization. For example, a startup accelerator might pitch its partnership with a major cloud provider as a way to "future-proof" an app, neglecting to mention that the resulting infrastructure dependency will make migration prohibitively expensive. A music distribution service might promote its integration with leading streaming platforms as a way to "maximize royalties," while structuring contracts so that the bulk of licensing revenue flows through intermediaries before reaching the artist.

Even the language used to describe these arrangements is part of the capture mechanism. Terms like "ecosystem," "marketplace," and "partner network" evoke images of open, collaborative spaces, yet in practice they often describe tightly controlled environments governed by centralized policies. Entry is contingent on compliance, and continued access is dependent on behavior that aligns with the controlling entity's objectives. This linguistic sleight of hand transforms rules into norms and restrictions into standards, making it difficult for creators to distinguish between what is truly necessary for professional growth and what merely serves the strategic interests of the gatekeepers.

The end result is an environment where creators are celebrated publicly for their innovation while privately constrained in how they can deploy it. Their intellectual property becomes an annuity for the platforms and their affiliates, a renewable resource harvested under the guise of partnership. And because the system is so deeply entrenched across multiple industries, the very act of seeking alternative routes often leads back to a different node within the same network. The web of players is expansive but not infinite; the names may change, but the structural imperatives remain constant.

In this sense, identifying the players is not simply a matter of naming corporations or institutions. It is about recognizing the systemic relationships that bind them together and understanding how those relationships are engineered to prioritize the continuous influx of intellectual property over the autonomy of its creators. The true power of the system lies not in any single entity's dominance, but in the collective architecture that ensures creators, regardless of their entry point, are funneled into an interdependent network from which exit is costly and rare. Only by mapping this architecture in its entirety can one begin to devise strategies for navigating it without succumbing to its most exploitative tendencies.

From Onboarding to Ensnarement

The process often begins in a haze of optimism, framed as the first real breakthrough in a creator's journey. The onboarding phase is carefully constructed to feel like a welcome into an exclusive circle, where the gatekeepers are not faceless institutions but attentive partners eager to nurture potential. Invitations arrive with the language of recognition, phrases that speak of "shared vision," "aligned values," and "mutual growth." The creator, often accustomed to years of self-reliance and incremental progress, is suddenly offered resources that promise to collapse timelines and multiply reach. Technical infrastructure, marketing exposure, and sometimes even direct financial incentives are presented as gestures of good faith, as though the platform's main priority is the creator's success. The reality, however, is that each of these initial benefits is designed not simply to empower, but to entwine.

The dependency begins subtly. Early-stage support frequently comes in the form of proprietary tools that integrate seamlessly into the creator's workflow. A developer might be given free credits for a cloud infrastructure that will later carry prohibitive migration costs. A musician may be provided with distribution channels that quickly become the primary source of audience engagement, even as the contractual terms governing those channels restrict the use of alternative outlets. A writer might be granted premium placement on a digital storefront in exchange for agreeing to exclusivity periods that extend far beyond the initial promotional window. Each act of assistance is coupled with a structural tether, terms of service, integration requirements, and revenue-sharing models that quietly anchor the creator to the platform's ecosystem.

The illusion of autonomy is maintained through the language of partnership. Platforms often emphasize that participation is voluntary and that creators retain ownership of their work. Yet buried within service agreements are clauses that grant the platform extensive usage rights, sometimes in perpetuity, over any derivative or adapted content created within its environment. The creator is assured that these provisions are "standard" and necessary to facilitate distribution or improve services. In practice, they become the legal scaffolding for long-term extraction, ensuring that even if the creator departs, the fruits of their labor remain monetizable within the platform's control.

Psychological levers are as important as technical and legal ones. Platforms understand that loyalty is not secured through coercion alone, but through a sense of belonging and validation. Recognition programs, badges, feature spotlights, "creator of the month" campaigns, are strategically deployed to foster emotional attachment. These gestures appear harmless, even flattering, but they create an implicit hierarchy in which the creator's status is tied to platform approval. Once this bond is established, the threat of losing visibility or access to special programs becomes a powerful tool of influence. The creator begins to shape decisions around what will maintain favor, often preemptively aligning their output with perceived platform preferences.

Economic levers further tighten the bind. Early revenue boosts, such as temporary increases in payout rates or bonuses for meeting production milestones, are designed to accelerate dependency. These incentives often taper off, replaced by baseline rates that seem modest in comparison but are still preferable to starting from scratch elsewhere. Meanwhile, the platform's control over discoverability, algorithmic promotion, search rankings, recommendation feeds, ensures that any significant deviation from the established partnership can result in a rapid and visible drop in audience engagement. This creates a feedback loop: the more the creator depends on the platform for income, the less they can risk actions that might reduce their visibility, and the more they adapt their creative process to fit within the confines of the platform's ecosystem.

Over time, the onboarding phase gives way to a more subtle, entrenched stage where support morphs into reliance and reliance becomes dependency. The early abundance of resources, promotional boosts, one-on-one guidance from account managers, priority access to beta features, gradually transitions into a pay-to-maintain structure. What was once provided without cost is now locked behind thresholds, premium tiers, or performance-based requirements. The creator is told they have "graduated" to a higher level of professionalism, and with that graduation comes the expectation to invest more into the platform, whether in the form of revenue share, advertising spend, or exclusivity commitments. This shift is rarely abrupt; it is carefully staged so that the creator, having reorganized their business model around the platform's tools and audience, sees compliance not as a loss of independence but as a natural evolution.

During this transition, the economic and psychological levers become tightly interwoven. The platform positions itself as both the source of opportunity and the arbiter of success, framing its algorithms as meritocratic mechanisms that reward quality and consistency. In truth, these algorithms function as finely tuned behavioral control systems, rewarding outputs that align with the platform's commercial objectives and deprioritizing those that do not. A subtle but powerful conditioning occurs: creators begin to preemptively tailor their work to what they believe the algorithm will favor, often at the expense of their own creative instincts. This form of anticipatory compliance deepens the bond, as the creator no longer merely reacts to platform incentives but actively shapes their identity and production schedule to fit them.

The emotional dimension of this conditioning cannot be overstated. For many, the period of initial onboarding represents the first time their work has reached a substantial audience or generated meaningful income. The memory of that breakthrough, of seeing follower counts spike, of receiving unexpected royalty deposits, of being publicly acknowledged, becomes a psychological anchor. Platforms leverage this memory in subtle ways, invoking it in personalized outreach, promotional materials, and performance dashboards. The implicit message is that this success was not solely the creator's doing, but the product of the partnership. The risk of leaving is framed not as the loss of a tool, but the loss of a relationship, a shared history, and a proven formula for achievement.

Economically, the architecture of dependency is reinforced by layered revenue entanglements. A creator's income on a given platform might be drawn from multiple streams, ad revenue, direct sales, tips or donations, sponsorship integrations, all of which are mediated, tracked, and disbursed by the platform. By integrating these streams into a single dashboard, the platform offers convenience while consolidating control. Payment terms, payout schedules, and eligibility criteria become levers that can be adjusted to reward compliance or penalize deviation. In extreme cases, funds can be withheld or accounts suspended under vague terms-of-service violations, serving as a deterrent not only to the individual but to the wider creator community observing the consequences.

Once the creator is fully embedded, the support mechanisms that once felt expansive are revealed to be boundaries. Opportunities for growth beyond the platform are subtly discouraged. Cross-platform promotion might be technically allowed but algorithmically downranked. Independent monetization avenues, selling directly to fans, offering premium content off-platform, may be restricted through exclusivity clauses or framed as breaches of policy. Even when no explicit prohibition exists, the sheer disparity in reach and engagement between platform-bound and off-platform activity disincentivizes any serious attempt at diversification. The platform becomes the primary, if not sole, channel for audience connection, and the cost of leaving is no longer measured only in lost income but in lost identity.

It is in this final phase of ensnarement that the true genius of the onboarding process is revealed. What began as an offer of empowerment has, through gradual shifts and carefully managed dependencies, transformed into a structure where the creator's economic survival, public visibility, and creative validation are all contingent on the maintenance of loyalty. The platform no longer needs to actively recruit or incentivize; the creator, conditioned by both fear of loss and hope of continued favor, sustains the relationship voluntarily. The velvet cage has been furnished to the point of comfort, and its walls, though invisible, are reinforced by every choice the creator makes to remain within it.

What makes this state so difficult to escape is that the boundaries are not enforced solely by the platform, they are internalized by the creator. The very metrics that once symbolized achievement become the chains that bind them. Follower counts, engagement rates, and monetization dashboards are not neutral indicators; they are engineered feedback loops that equate continued loyalty with continued worth. The creator begins to monitor these metrics obsessively, adjusting their schedule, content style, and even personal expression to maintain upward trends. In this way, the platform succeeds in outsourcing its enforcement mechanisms to the creator's own self-perception. The fear of seeing those numbers decline can be as powerful a deterrent as any contractual restriction.

Meanwhile, the economic entanglement deepens. As the creator invests more time and resources into optimizing their platform presence, purchasing advertising credits, upgrading to premium analytics, hiring contractors to produce platform-friendly assets, the sunk cost grows heavier. The psychological principle of loss aversion ensures that the idea of abandoning this investment becomes increasingly unthinkable. Even if another platform or independent model offers better long-term terms, the creator hesitates, unwilling to risk the short-term disruption that could come from stepping outside the established system. This hesitation is precisely the point: the onboarding process was never about providing temporary aid; it was about designing a dependency so resilient that the creator will maintain it themselves.

Platforms also employ scarcity and exclusivity to fortify loyalty. Special programs, early access to new features, invitations to private creator summits, inclusion in high-profile promotional campaigns, are offered to a select few, often with opaque criteria. This selective recognition fosters a competitive environment among creators, where maintaining one's privileged status becomes a goal in itself. The possibility of losing that access, even for reasons unrelated to performance, acts as a quiet yet potent behavioral control. It encourages conformity, reduces public dissent, and keeps creators vying for the platform's attention rather than questioning its terms.

This environment is further reinforced by the social dimension of creator communities. Platforms often cultivate networks of interconnected creators, encouraging collaborations, cross-promotions, and shared branding opportunities, all within the confines of the platform's infrastructure. Over time, a creator's professional relationships, industry reputation, and support networks are largely populated by others who are equally bound to the same system. This makes the idea of leaving not only an economic risk but a social rupture. To walk away is to distance oneself from collaborators, mentors, and peers who remain invested in the ecosystem. It is to forfeit the sense of belonging that was carefully nurtured from the moment of onboarding.

Perhaps the most sophisticated aspect of this ensnarement is that it often masquerades as mutual success. The platform can point to tangible metrics, the creator's increased audience, higher earnings compared to their pre-onboarding days, greater visibility in their field, as evidence of a thriving partnership. And for the creator, these gains are real, making it harder to frame the relationship as exploitative. Yet the critical distinction lies in proportionality: while the creator's gains may be significant in absolute terms, they are often dwarfed by the value their intellectual property generates for the platform. This imbalance is obscured by the constant framing of benefits in personal terms, diverting attention from the structural asymmetry that underpins them.

By the time the creator fully recognizes the depth of their dependency, the escape routes have narrowed. The infrastructure they rely on is owned and operated by the very entity they might wish to leave. The audience they have cultivated is algorithmically tethered to the platform's distribution channels. The monetization mechanisms they use are governed by terms that can be altered unilaterally. In practical terms, departure would require rebuilding not just an income stream but an entire operational framework, often without access to the historical data, audience insights, or creative assets that would make such a transition viable.

Thus, the onboarding process achieves its ultimate purpose: not merely to attract talent, but to bind it so intricately to the platform's systems, through technical integrations, contractual obligations, economic incentives, and psychological conditioning, that the prospect of leaving becomes indistinguishable from the prospect of starting over entirely. It is a design of quiet inevitability, where the path from invitation to captivity is so gradual, so cloaked in the language of partnership and progress, that by the time the creator understands its true contours, they have already been woven into the fabric of the machine they once believed was lifting them up.

The Profiling Process

The profiling of high-value creators begins long before most realize they are even on a platform's radar. In the modern digital economy, where nearly every click, view, and engagement is recorded, the first layer of identification happens passively. Algorithms ingest vast amounts of behavioral data, not just the obvious metrics like follower count or video views, but granular patterns such as engagement velocity, cross-platform resonance, audience retention curves, and even the sentiment expressed in comment threads. A creator whose work consistently triggers spikes in these indicators is quietly marked for closer observation. This initial flagging is not the result of human curation in most cases; it emerges from machine learning models trained to detect anomalies of value, content or personalities that outperform expected baselines in ways that could translate into sustained monetization.

Once the signal is strong enough, it passes into the next stage, where algorithmic detection merges with network analysis. Here, platforms examine the creator's position in the social graph: Who engages with them? How interconnected are their followers? Do they bridge distinct audience clusters that the platform would otherwise struggle to connect? A creator who can serve as a node linking disparate communities is far more valuable than one who dominates a single niche, because they can amplify reach across multiple demographics without additional marketing spend. The models assign weightings to these relational factors, effectively ranking creators not just by the content they produce but by their structural position in the digital ecosystem.

The public-facing veneer of this process is often dressed up as "creator analytics" or "insight dashboards," where the creator can see simplified versions of their own data. What is hidden from view is the deeper layer of profiling, the invisible "talent scoring" systems that integrate performance, network position, content characteristics, and behavioral compliance into a single internal rating. These scores are dynamic, recalculated in real time as the creator's metrics shift. They are rarely disclosed, yet they dictate which creators are surfaced to partnership teams, invited to exclusive programs, or granted early access to monetization features.

While platforms present these opportunities as merit-based, the scoring models are built to align with corporate priorities. A creator producing content that drives high ad engagement, for instance, might be weighted more heavily than one with similar reach whose audience is less responsive to commercial prompts. Likewise, a creator who reliably adheres to platform policies and avoids controversy, regardless of artistic merit, may receive a higher reliability index, making them a safer long-term investment. The score is not a neutral reflection of "talent" in any broad sense; it is an index of predicted profitability and risk.

The sophistication of these systems lies in their predictive capacity. By correlating early-stage signals from past breakout stars, platforms can identify creators with similar trajectories before they peak. This allows for intervention at a point when the creator is still accessible, eager for recognition, and more likely to accept onboarding offers without the leverage that fame might later bring. The invitation to a beta program or a "creator accelerator" is not random generosity; it is a calculated move timed to coincide with the moment when the creator's dependency can be most effectively cultivated.

Once a creator enters this predictive spotlight, the platform's outreach becomes more deliberate and tailored. Internal partnership teams, armed with the data from these hidden scoring systems, approach with offers that seem personalized because they are, down to the frequency of posting, preferred content formats, and even the times of day their audience is most active. This is not guesswork; it is the product of behavioral modeling that has mapped the creator's patterns with precision. What appears to the creator as an organic recognition of their unique strengths is, in reality, a confirmation that their metrics align neatly with a profitable archetype the platform already understands how to monetize.

In parallel, network influence is measured and leveraged. Platforms look beyond the creator's direct audience to assess their secondary and tertiary reach: who shares their content, which influencers or micro-communities engage with it, and how quickly it propagates through the ecosystem. A high-value creator is rarely an island; they are a hub whose content can ignite a cascade of interactions across different segments of the user base. This ripple effect is quantifiable, and it becomes a key factor in their internal valuation. The more efficiently a creator can mobilize others, willingly or unwittingly, the higher their score climbs.

The invisible scoring doesn't stop at influence and reach. It quietly tracks behavioral compliance. This includes the creator's responsiveness to platform nudges, such as prompts to adopt new features, participate in seasonal campaigns, or adjust content to fit trending formats. Those who demonstrate flexibility and quick uptake are tagged as "cooperative," a quality that increases their attractiveness to partnership managers. Conversely, creators who ignore or resist these nudges may see their visibility subtly throttled, not necessarily as punishment, but as an allocation of resources toward "better-aligned" talent.

In many cases, these scores are cross-referenced with third-party datasets obtained through advertising partnerships, data brokers, or affiliated services. This means a creator's off-platform activities, public speaking engagements, brand collaborations, media coverage, can be folded into their talent profile. The picture that emerges is not a snapshot but a living dossier, one that evolves with every post, click, and interaction. This dossier determines not only who is fast-tracked for growth opportunities but also the sequence in which those opportunities are offered, ensuring that the most pliable and profitable candidates are prioritized.

The creator, of course, is largely unaware of this parallel system of evaluation. The visible markers of success, feature placements, sudden boosts in algorithmic visibility, invitations to join closed groups, feel like rewards for effort, when they are in fact milestones in a premeditated capture process. The danger lies in the asymmetry of awareness: the platform knows exactly where the creator stands in its internal rankings and how that ranking is trending, while the creator sees only the outcomes, never the calculus that produced them.

This disparity allows the platform to shape the creator's sense of agency. By selectively surfacing opportunities that align with both the creator's aspirations and the platform's commercial goals, it encourages the belief that continued compliance will lead to sustained elevation. The reality is that the scoring system is designed to maximize return on investment, and once a creator's profitability plateaus, or their willingness to conform diminishes, they can be quietly deprioritized in favor of the next rising prospect already identified by the algorithm.

Over time, this silent ranking infrastructure becomes the unseen hand that guides a creator's entire trajectory. The pace of their growth, the scope of their visibility, and even the types of brands or collaborators that approach them are all downstream effects of a score they cannot see and cannot challenge. It is an architecture of influence that functions without overt coercion; instead, it rewards alignment so predictably that creators often self-regulate their behavior to maintain or improve their standing. They internalize the platform's values as professional best practices, believing they are simply being "strategic" when, in truth, they are conforming to an invisible template optimized for the platform's profitability.

One of the most effective tools within this system is the blending of algorithmic judgment with human validation. Once a creator's score crosses a certain threshold, they may receive outreach from a dedicated partner manager or community liaison. This human contact is not a replacement for algorithmic control but a complement to it, designed to strengthen the emotional bond between creator and platform. The partner manager becomes a conduit of both encouragement and subtle pressure, reinforcing the idea that the creator's continued success depends on staying within certain parameters. Because this relationship feels personal, advice and feedback are received with more trust, even when they serve the platform's interests more than the creator's independence.

The profiling process is also adept at anticipating moments of vulnerability. By monitoring fluctuations in engagement, output, or sentiment, the system can predict when a creator might be more receptive to intervention, perhaps after a period of declining views or a public controversy. Strategic offers often appear at these inflection points: temporary boosts in visibility, invitations to participate in high-profile campaigns, or exclusive monetization tools. These are framed as lifelines, but they function as renewed binding contracts, reestablishing the platform's central role in the creator's survival and growth.

Invisible "talent scoring" further extends its reach by operating across interconnected platforms within the same corporate family or partner network. A creator's activity on one service can influence their ranking on another, creating an interdependent web of evaluations. This cross-platform profiling magnifies the power imbalance: a drop in compliance or profitability in one environment can ripple across others, effectively lowering the creator's standing in multiple arenas at once. Conversely, strong performance in one space can fast-track them into opportunities elsewhere, reinforcing the sense that the safest and most lucrative path is full integration into the platform's broader ecosystem.

The sophistication of these systems lies not only in their predictive power but in their opacity. Creators rarely know which specific actions raised or lowered their internal ranking, and this uncertainty itself becomes a control mechanism. In the absence of clear feedback, they are incentivized to err on the side of over-compliance, avoiding risks, experimenting less, and increasingly centering their creative process around what they believe the system wants. In this way, the profiling process not only identifies high-value creators but actively shapes them into more dependable, less unpredictable assets.

By the time a creator recognizes that their career has been scaffolded around an invisible score, the dependencies are deeply entrenched. Their audience reach, monetization potential, and professional network are all conditioned on maintaining a favorable ranking within a system they neither designed nor consented to in any meaningful way. The profiling process, therefore, is not merely a tool of observation; it is a mechanism of quiet assimilation, turning independent innovators into predictable components of a larger commercial machine, valued, yes, but always valued on terms that are not their own.

Case snapshots

One of the clearest illustrations of the capture process comes from the experience of an independent filmmaker whose breakout short went viral on multiple social platforms in the span of a week. Until that moment, his work had been self-funded and distributed directly to niche online communities. Within days of the surge, he was approached by representatives from a major streaming service with promises of "creative freedom" and "access to unparalleled resources." The onboarding was swift, technical assistance, grants, and promotional placement on the platform's front page. But embedded in the agreement were clauses requiring all future projects, regardless of genre or scale, to be pitched first to the platform, with an exclusivity window that stretched for years. The filmmaker initially saw this as a fair trade for exposure and financial stability. It wasn't until he tried to produce a side project for a nonprofit organization, only to be told the rights would automatically belong to the platform, that he recognized how thoroughly his independence had been absorbed into their infrastructure. By then, his audience, revenue streams, and even his post-production workflow were so intertwined with the platform's systems that extricating himself would mean dismantling the very career their partnership had built.

Another case emerged in the gaming world, where a small development team built a cult following through a series of innovative browser-based titles. Their distinctive style and high player retention rates triggered a platform's internal "talent scoring" system, leading to an offer of integration into the platform's premium game hub. The arrangement promised server support, global distribution, and marketing campaigns that dwarfed what the team could achieve on its own. In exchange, the platform took a majority share of in-game monetization and inserted contractual language giving it right of first refusal on all future titles. The developers accepted, believing the partnership would allow them to focus solely on creativity. Over time, however, they found themselves adjusting game mechanics to meet algorithmic preferences, shorter play sessions, more ad breaks, in-game purchases tied to trending microtransactions. By the time their original fan base noticed the shift in tone and began drifting away, the studio was locked into an ecosystem where their visibility, income, and development tools were all contingent on maintaining the platform's preferred design patterns.

A different form of capture played out in the publishing industry, where an independent author built a large following by releasing serialized fiction directly through her website and email list. A dominant e-book retailer reached out with an offer to feature her work prominently if she agreed to make it exclusive to their marketplace. The incentive was immediate: a sharp increase in sales, access to the retailer's promotional tools, and invitations to speak at industry events. But over the next two years, the retailer's algorithm began to favor shorter works with rapid release schedules, pressuring her to alter her writing process. When she attempted to experiment with a slower, more ambitious project, its visibility plummeted, and the loss in income made it financially unfeasible to continue. By that point, her direct sales channels had withered, her email list engagement had declined, and her readership was conditioned to find her work only through the retailer's interface.

Even visual artists are not immune to such structures. One painter gained significant traction on a popular social media platform known for image sharing. After being identified by the platform's network analysis as a high-engagement creator, she was offered a partnership that included sponsored posts, priority placement in explore feeds, and access to beta tools for selling prints. The partnership contract also allowed the platform to use her images in promotional material without additional compensation. Over time, she realized that her posts which did not feature products or align with certain visual trends received far less reach, creating a subtle but powerful pressure to adjust her style. When she tried to diversify her online presence by promoting a gallery show through another platform, her engagement on the original network dropped sharply, whether by algorithmic design or coincidence, she could not determine. The chilling effect was the same: she returned to posting in ways that maximized platform approval rather than her own artistic priorities.

Across these stories, the pattern is unmistakable. The initial approach is framed as a recognition of unique value, backed by resources that promise to elevate the creator's work. But the agreements, whether explicit in legal terms or implicit in algorithmic reinforcement, reshape the creator's incentives, creative decisions, and public identity until departure feels synonymous with erasure. By the time these individuals recognize the extent of their integration into the platform's systems, the cost of leaving is not just professional inconvenience; it is the dismantling of the very structure that their career now rests upon.

In the music industry, a striking example came from a self-produced singer-songwriter who had cultivated a modest but fiercely loyal following through live performances and direct-to-fan sales. A major streaming platform identified her as a promising "emerging artist" based on audience growth velocity and the unusually high completion rate of her tracks. She was offered placement on curated playlists with millions of followers, coupled with exclusive access to the platform's in-house recording facilities. The visibility spike was immediate and intoxicating; her monthly listener count increased tenfold in under two months. Yet the agreement included non-negotiable exclusivity terms for all new releases, with the platform controlling the timeline of song drops to optimize their own content cycles. As she became more enmeshed, she found herself tailoring her songwriting to align with the playlist algorithms, shorter intros, hook placement within the first fifteen seconds, and lyrical content that matched the mood tags favored by the platform's categorization system. When she attempted to release an experimental track independently, she learned that her contract effectively barred her from distributing it elsewhere for a period of 18 months. By then, her audience had been conditioned to engage with her music solely through the platform, severing the direct artist-fan relationship she had worked years to build.

A similar dynamic unfolded in the world of online education. A pair of educators had created a series of high-quality, independently hosted video courses in advanced programming, attracting a small but lucrative subscriber base. They caught the attention of a major e-learning marketplace, which offered them front-page placement, marketing support, and a revenue-sharing agreement that, at first glance, seemed generous. The catch was a pricing model dictated by the marketplace, which undercut their original rates to align with platform-wide sales strategies. While enrollments skyrocketed, the per-student revenue dropped to a fraction of what they had previously earned, forcing them to produce more courses at a faster pace just to maintain income. The marketplace also claimed perpetual licensing rights to any course uploaded, preventing the educators from repackaging or reselling their own content elsewhere. When they eventually attempted to leave, they found that their brand had become inseparable from the platform's identity, students searching for their names were directed to the marketplace's pages, not their independent site.

In the sphere of social commentary and journalism, a digital columnist gained notoriety for incisive, long-form analysis shared via a personal blog and newsletter. His growing readership attracted a large publishing platform that promised higher visibility, built-in monetization, and a share of ad revenue. Within a year, his articles were featured prominently on the platform's trending pages, and his readership tripled. However, the platform's recommendation engine began nudging him toward more frequent, shorter pieces optimized for click-through rates. Stories with nuanced or unpopular angles were deprioritized, limiting their reach and undermining the very depth that had drawn readers to him initially. The platform also inserted its own advertisements and partnerships into his work, without giving him veto power over brands or messages. When he explored moving back to independent publishing, he discovered that the majority of his subscribers had subscribed through the platform's internal system, meaning he could not contact them directly without paying steep fees to "export" his own audience.

A final case involved an innovative fashion designer who had gained momentum through a small but influential circle of stylists and boutique owners. An e-commerce giant approached her with the promise of a dedicated storefront, targeted advertising, and international distribution. The onboarding process was seamless, and sales surged as her designs were recommended alongside established luxury brands. Yet the contract required her to maintain a certain inventory turnover, pressuring her to produce designs that aligned with fast-fashion cycles rather than her slower, sustainable model. The platform's analytics team began sending her "performance suggestions," which in practice meant adapting her styles to match trending search terms and high-conversion color palettes. When she resisted, her product placements gradually slipped from premium recommendations to lower-visibility categories. Within two years, the bulk of her sales came from the platform's ecosystem, and her independent retail relationships had atrophied. Attempting to return to those channels required reintroducing herself to buyers who now associated her brand entirely with the e-commerce giant's mass-market positioning.

These stories, spread across industries as varied as film, gaming, literature, music, education, journalism, and fashion, share a common arc. Each creator began with an independent vision and direct connection to their audience. Each was approached by a larger entity that offered resources, exposure, and legitimacy. And in each case, the conditions of that partnership, whether expressed in contractual terms, algorithmic incentives, or cultural pressures, gradually reoriented the creator's work toward the platform's objectives. By the time the full scope of dependency became clear, the creator's professional infrastructure, audience relationships, and revenue mechanisms were so bound to the platform that departure meant starting over from near zero.

In the realm of independent game streaming, a charismatic broadcaster built a following of several hundred thousand viewers through a combination of skillful gameplay and an unfiltered, engaging personality. Their growth caught the attention of one of the largest live-streaming platforms, which extended a lucrative exclusivity deal. The package included featured placement on the homepage, technical support for high-quality streaming, and early access to new monetization tools. Initially, the agreement felt like a dream realized, donations surged, sponsorship offers multiplied, and their subscriber count doubled. But the contract stipulated minimum streaming hours per month and penalized prolonged absences, effectively chaining the creator to a schedule dictated by the platform's content supply needs rather than their own creative rhythm or personal well-being. Over time, algorithmic recommendations increasingly favored "safe" content and heavily moderated language, nudging the streamer to dilute the edge that had originally set them apart. When mental fatigue eventually forced them to take a break, the algorithm swiftly deprioritized their channel. Upon returning, their audience engagement had cratered, and rebuilding outside the platform was nearly impossible because the subscriber contact data belonged entirely to the company.

In the visual media production space, a small documentary collective focused on social justice issues found themselves in a similar bind. Their work garnered attention at film festivals and across grassroots networks, prompting a popular video hosting site to offer sponsorship for a series on global human rights. The hosting site provided funding, promotion, and analytics support, but required that all distribution take place on their platform first, with a six-month delay before any content could be released elsewhere. During the partnership, the collective was encouraged to adjust topics to maximize advertiser comfort and viewer retention metrics, resulting in more generalized, less confrontational storytelling. Their internal analytics showed that while view counts were up, deep engagement, measured by shares within activist networks and audience follow-up actions, was falling. When the series concluded, the platform opted not to renew funding, leaving the collective with diminished independence, a fractured audience, and limited rights over their most impactful work.

The world of app development offers another telling example. A two-person startup created a simple productivity app that quickly rose to the top of download charts due to its elegant design and intuitive features. A major app store operator approached them with an offer to feature the app prominently in exchange for adopting the platform's in-app purchase framework and integrating with its proprietary cloud service. The resulting exposure generated a flood of new users, but revenue splits heavily favored the store operator, and integration requirements made the app dependent on APIs that could change without notice. When the developers considered adding features that would work better with third-party services, they found that doing so would breach their agreement. Eventually, an API change imposed by the store forced them to overhaul their backend architecture, consuming months of work just to maintain compatibility. By then, their brand identity was so tied to the app store's ecosystem that launching on a competing platform would have been akin to starting from scratch.

Even in the sphere of independent podcasting, the capture dynamic plays out with familiar precision. A pair of investigative journalists built a loyal audience through a crowdfunded series tackling local political corruption. They were approached by a major audio streaming platform that promised studio resources, marketing support, and integration into a premium subscription tier. The agreement required exclusivity, with all episodes locked behind the platform's paywall after a short public window. While initial revenue exceeded their crowdfunding income, listener feedback revealed frustration over restricted access, and growth slowed as casual listeners were unwilling to pay for a single show. Additionally, the platform's editorial team began suggesting story angles that would better align with trending topics and advertiser interests, gradually steering the journalists away from the slow, meticulous investigations that had built their credibility. By the time they considered returning to open distribution, their listener data and mailing list were locked behind the platform's proprietary analytics system, effectively cutting them off from the very community that had launched their work.

Each of these examples demonstrates how the capture process is not confined to any one industry or creative medium. The specifics of the hook vary, playlist placement for musicians, front-page visibility for app developers, homepage features for streamers, or premium tier inclusion for podcasters, but the structure is consistent. The entry point is always framed as opportunity, the middle stage as partnership, and the outcome as dependency. By the time creators in these scenarios fully recognize the asymmetry in their relationships, the mechanisms of control are already deeply embedded, not just in contracts, but in the technological frameworks, audience behaviors, and professional ecosystems that now define their careers.

Chapter 2, The Marking of the Golden Goose

Early Signals

The earliest indicators that a creator has entered the periphery of a platform's high-value watchlist rarely appear as threats. They arrive dressed in the garments of recognition, small but deliberate gestures that signal the platform has taken notice. A message from a community moderator congratulating the creator on a milestone, an unexpected follow from an official corporate account, or a personalized invitation to "chat about opportunities" all seem innocuous, even flattering. These moments are engineered to plant the seed of reciprocity, drawing the creator into a conversational rhythm where the platform's representatives become familiar presences rather than abstract gatekeepers. It is in these exchanges, often couched in warmth and collegiality, that the first threads of a more permanent relationship are woven.

This increased attention is rarely random. Platforms monitor their ecosystems with an intensity far exceeding public perception, using algorithmic flagging systems to identify individuals whose metrics, engagement velocity, retention rates, or network reach, suggest outsized potential. Once identified, the creator is often passed into the hands of a dedicated liaison whose role is part talent scout, part brand ambassador, and part compliance officer. These representatives are trained not only to highlight the platform's strengths but also to frame the creator's recent success as inseparable from the platform's ecosystem. In doing so, they subtly establish a narrative in which continued growth is implicitly tied to deeper integration.

One of the most effective tools in this early phase is the invitation to a "special program" or "exclusive beta." These initiatives are positioned as opportunities to gain early access to features, to network with other rising talents, or to receive direct feedback from product teams. In practice, they serve multiple functions. They allow the platform to collect granular data on how top-performing creators interact with new tools, to fine-tune algorithms using real-world content, and to normalize the idea that the creator's work is best shaped in collaboration with, if not outright guided by, the platform's strategic objectives.

The exclusivity of these programs is a powerful psychological lever. Creators are reminded that only a small fraction of applicants are accepted, that they are among "the chosen few." This scarcity effect amplifies the perceived value of the opportunity and fosters a sense of obligation to reciprocate the favor. What goes unspoken is that these programs also create a control environment in which the platform can experiment with feature rollouts, policy enforcement, and content guidelines on a compliant, high-value subset of users. Those who adapt quickly and produce results that align with the platform's internal metrics are rewarded with more visibility, more resources, and more personalized outreach. Those who do not are quietly cycled out, their early momentum dissipating without public explanation.

At this stage, nothing overtly coercive has taken place. The creator retains full legal rights to their work, can technically leave the platform at any time, and may even see a genuine boost in their reach and income. But beneath the surface, dependencies are already forming. The creator's workflow begins to accommodate the platform's tools, their content evolves to take advantage of beta features, and their audience starts to associate their brand with the platform's identity. Each of these changes is incremental, almost imperceptible in isolation, but together they create an environment in which stepping away begins to feel like relinquishing progress.

The transformation deepens as these touchpoints become more frequent and more personalized. The platform representative evolves from a distant figure into something closer to a trusted advisor, checking in regularly to offer "insider" tips on optimizing content or increasing audience engagement. These suggestions are framed as neutral best practices, but they are often subtly aligned with the platform's commercial goals, encouraging formats that boost ad impressions, promoting topics that match seasonal marketing pushes, or prioritizing posting times that fill gaps in the platform's content flow. The creator, eager to maintain momentum and appreciative of the guidance, incorporates these adjustments without fully considering how much they are shaping their creative direction to fit a blueprint they did not design.

Invitations to exclusive betas often follow a similar arc. At first, the allure lies in the promise of early access, a chance to test tools or features before they are available to the general public. For a creator hungry to innovate, the appeal is obvious: they can experiment ahead of competitors, differentiate their output, and perhaps even influence the development process with their feedback. But these betas are rarely neutral playgrounds. Participation often requires the use of proprietary infrastructure, embedding the creator's workflow more deeply into the platform's technical ecosystem. Content produced in these test environments may be stored, analyzed, and repurposed by the platform under expansive licensing terms buried in the agreement. And because these features are not yet public, the creator has no realistic way to transfer the workflows or formats they develop during the beta into another ecosystem without losing the unique advantages they have cultivated.

The exclusivity also has a social dimension. Beta participants are often grouped into private forums or invited to closed networking events where they can connect with other high-performing creators. These spaces can be energizing, full of peer recognition, shared problem-solving, and a sense of being part of a select community. Yet they also serve to concentrate valuable social capital within the platform's boundaries. The relationships formed in these settings, whether with peers or with platform staff, are mediated entirely through the host's channels. Should the creator ever consider leaving, they risk losing not only their access to special tools but also the professional network they have built inside this curated environment.

The cumulative effect of these early signals is to create a feedback loop of validation and dependency. The creator sees tangible gains, more followers, higher engagement, new opportunities, and associates them with the increased contact and privileges granted by the platform. The platform, in turn, uses the creator's compliance and performance data to refine its profiling systems, ensuring that only those who align well with its objectives continue to receive such treatment. This self-reinforcing cycle makes it increasingly difficult for the creator to separate the value of their own work from the value conferred by their association with the platform.

By the time the creator begins to recognize patterns in the guidance they are receiving or notices how their audience engagement fluctuates in tandem with their participation in special programs, the lines between independence and integration have already blurred. They may still feel free, still believe they are charting their own course, but the architecture of subtle control is firmly in place, ready to tighten whenever the platform decides the moment is right.

As the relationship matures, the pace and precision of these signals increase, creating an almost imperceptible rhythm that draws the creator deeper into alignment. A well-timed message from a representative coincides with a sudden uptick in algorithmic visibility, reinforcing the idea that engagement with the platform's guidance produces immediate and measurable results. The next invitation arrives not in the form of a generic mass email but as a carefully crafted offer tailored to the creator's niche, referencing specific content they've produced and suggesting how the platform's latest initiative is a "perfect fit" for their style. This personalization cultivates a sense of being understood on an intimate level, which in turn encourages trust and cooperation.

Over time, the creator begins to anticipate these touchpoints, subtly orienting their creative decisions toward attracting more of them. If certain types of posts tend to trigger outreach from the platform, those posts become more frequent. If participation in a beta coincides with a spike in follower growth, the creator is more likely to prioritize beta-related content over independent experiments. The early signals, once simply flattering, become behavioral cues that shape output, pacing, and presentation.

The strategic brilliance of this stage is that it requires no overt instruction. The creator believes they are making independent choices, but those choices are guided by the memory of rewards, an increase in views, a new connection made during an exclusive event, a feature on the platform's homepage, each one linked to compliance with subtle platform directives. By the time they realize how much of their trajectory has been influenced by these cues, their professional identity has been recalibrated to thrive within the platform's framework.

Even the rare moments of withdrawal, periods when a creator is not actively engaged in a program or beta, are leveraged as part of the long game. A lull in communication from platform representatives can coincide with a drop in visibility, reinforcing the unspoken connection between engagement and reward. When contact resumes, often accompanied by a fresh opportunity, the return of attention feels like the lifting of a shadow, making the creator more likely to accept without hesitation. This oscillation of scarcity and abundance ensures that the platform retains psychological leverage even without direct intervention.

The earliest signals, then, are not mere gestures of goodwill. They are the opening moves in a longer strategy of capture, carefully calibrated to appear as organic recognition, yet rooted in systems designed to measure, predict, and shape creator behavior. Each congratulatory message, each exclusive invitation, and each period of curated silence is part of a sequence that builds dependency before the creator even suspects they are being positioned for long-term integration. By the time the reality becomes clear, those initial signs of interest, once seen as lucky breaks or hard-earned validation, reveal themselves as the first threads of a net that has been tightening from the very start.

Quiet Containment

Quiet containment often begins without a single notification, warning, or visible signal to the creator. It is a shift in the invisible architecture around them, a recalibration of how their content is surfaced, recommended, or monetized, subtle enough that it can be mistaken for organic fluctuation. The absence of an explicit trigger is what makes it so effective; there is no policy violation to dispute, no formal strike to contest, only a gradual narrowing of reach that quietly resets the boundaries of what is possible. This is the point where a creator, once encouraged to grow without limit, is quietly placed into a controlled environment calibrated for the platform's interests rather than their own.

The most common lever in this phase is the manipulation of discoverability. Platforms possess vast, complex ranking systems that determine who sees what, when, and under which circumstances. By altering the weight given to certain signals, engagement velocity, historical retention, audience interaction ratios, the system can shift a creator's visibility without removing their content outright. Instead of being prominently recommended to new audiences, their work is shown primarily to existing followers, or worse, to a fraction of them. The change can be imperceptible at first, masked by normal ebb and flow in audience activity. But over weeks and months, the creator notices that the serendipitous influx of new viewers slows, the growth curve flattens, and even long-time followers start to comment that they haven't seen posts in their feed for some time.

This quiet throttling is not always punitive. Often, it is strategic resource allocation. Platforms have finite attention to distribute, and they deploy it toward creators and content types that are currently most aligned with their commercial priorities. In practical terms, this means that when a creator's style, subject matter, or audience demographics no longer align with those priorities, the algorithm reallocates exposure elsewhere. No message is sent to the creator to explain this shift; to do so would risk revealing the extent of algorithmic control and provoke demands for transparency. Instead, the absence of opportunity is left to be interpreted as a plateau in relevance or a lapse in quality.

Monetization controls are similarly opaque. A creator may notice that their ad revenue per view has declined or that certain pieces of content, identical in format to earlier, well-performing work, are deemed ineligible for full monetization. In some cases, these restrictions are justified with vague terms like "advertiser suitability" or "brand safety," but the actual decision-making process is buried within automated systems that apply context-blind filters. The result is a chilling effect: creators begin to avoid topics, formats, or creative risks that might trigger reduced payouts, effectively policing themselves into conformity with the platform's unstated preferences.

What makes quiet containment particularly insidious is that it can be deployed selectively and dynamically. A creator may be throttled in one content category while performing normally in another, making it difficult to detect a pattern. Or they may experience temporary boosts, brief surges in exposure that seem to confirm their relevance, only to return to suppressed visibility shortly afterward. This intermittent reinforcement keeps them engaged, chasing the conditions that produced the spike, even when those conditions are opaque or constantly shifting. It also obscures the fact that the platform, not the audience, is the primary determinant of their reach.

The mechanics of algorithmic manipulation are often invisible even to those within the company, compartmentalized among engineering teams, data scientists, and product managers. Decisions about how to weight certain engagement signals, when to introduce friction into a content distribution pathway, or how to segment audiences for exposure are made incrementally and tested through constant A/B experimentation. From the creator's perspective, the changes feel organic, as though they are competing in an open market of attention. In reality, they are operating within a managed economy where supply and demand are balanced not by audience choice but by the platform's revenue model.

Over time, containment changes the nature of the relationship between creator and platform. In the growth phase, the platform appears as a partner, amplifying reach and rewarding engagement. In containment, the platform becomes a gatekeeper whose approval is granted conditionally, often through unexplained patterns of visibility and suppression. The creator, unable to access the logic behind these patterns, begins to make creative decisions in a fog of second-guessing, prioritizing safety over innovation. This erosion of creative autonomy is rarely acknowledged by the platform, yet it is one of the most potent tools for ensuring that high-value creators remain productive without becoming unpredictable.

The containment phase also exploits the creator's own investment in the system. By the time these hidden limits take effect, most have already adapted their workflow, schedule, and even personal brand identity to fit the contours of the platform's algorithms. They have invested in platform-specific equipment, learned its preferred formats, and built an audience whose habits are shaped by the platform's interface. These sunk costs weigh heavily in every decision. When visibility dips or monetization tightens, the rational impulse is not to leave, but to try harder within the same system, producing more content, engaging more often, and spending more money on platform-sanctioned promotional tools in the hope of reversing the trend. Each of these efforts deepens the dependency the containment phase is designed to cement.

Algorithmic exposure becomes the most powerful point of leverage. A subtle adjustment to ranking logic can determine whether a video appears on a homepage carousel, whether a post is recommended to non-followers, or whether it remains buried under an endless scroll of competing content. These micro-adjustments are invisible to the user but decisive in shaping outcomes. The platform can modulate this exposure as finely as a dimmer switch, raising it just enough to maintain the creator's belief in attainable growth, then lowering it to keep that growth within predictable boundaries. The effect is a controlled climate, neither so cold that the creator abandons the platform, nor so warm that they outgrow its authority.

Monetization throttles operate in parallel. Some creators experience unexplained delays in payment processing or see a growing percentage of their content labeled as "limited monetization" without clear justification. Others notice that ad placements become less frequent or shift toward lower-value advertisers, reducing revenue per impression. Because these changes are often rolled out quietly, under the guise of "system updates" or "policy enhancements," they are difficult to contest. The platform can always point to generalized rules, while avoiding any acknowledgment that individual accounts are being managed differently behind the scenes.

This combination of reach limitation and monetization control fosters a quiet compliance. The creator learns, often unconsciously, which topics, formats, or tones keep them in the algorithm's good graces. They stop experimenting in ways that might jeopardize that fragile stability, narrowing their own range of expression until it aligns almost perfectly with the platform's commercial objectives. Over time, their creative output becomes optimized not for audience connection or artistic fulfillment, but for algorithmic compatibility, a compatibility that the platform itself defines, measures, and changes at will.

In this way, quiet containment achieves what blunt censorship or overt contractual restrictions could not. It keeps creators within an acceptable range of profitability and predictability without provoking the backlash that open suppression would invite. The architecture of control remains hidden behind the façade of an open, merit-based system. From the outside, it appears that the creator is still free, still active, still thriving. Only by tracing the slow arc of their trajectory, the gradual flattening of growth curves, the subtle narrowing of their thematic range, the quiet erosion of their independence, does the reality of containment come into view.

Data Harvesting

Data harvesting begins the moment a creator steps onto a platform, though in its early stages it is framed as nothing more than the collection of innocuous "engagement metrics" to improve user experience. In reality, the scope and precision of behavioral analytics extend far beyond the data points the creator is shown. Every click, pause, rewatch, hover, and scroll becomes part of an expanding behavioral profile, cross-referenced with demographic markers, device data, geographic location, and even inferred personality traits. The stated purpose is optimization, helping the creator "better understand their audience", but the deeper function is twofold: to predict the creator's next move and to shape it.

The carrot in this equation is the feedback loop of flattering insights. A creator dashboard might highlight which pieces performed best, the optimal time to post, or the audience segments that responded most strongly. Presented as empowering intelligence, this data becomes addictive. Creators log in not only to publish, but to read their own numbers as if scanning a mirror that reflects their worth. They learn to chase the spikes and avoid the dips, aligning their behavior ever more closely with what the analytics suggest will yield positive results. Over time, their creative decisions are less about artistic vision or independent strategy and more about maintaining the metrics that keep them in the algorithm's good graces.

The stick emerges in the opaque penalties that follow deviation. A creator who experiments outside their proven formula may notice a sudden drop in reach, with no accompanying explanation. This is not always punitive in the formal sense; it is the algorithm recalibrating its "confidence" in the creator's predictability. Yet from the creator's perspective, the message is unmistakable: divergence from the data-validated path carries risk. The absence of transparency turns every underperforming post into a lesson in self-restraint, pushing the creator toward safer, more platform-aligned output.

What makes behavioral analytics uniquely potent as a control mechanism is that the platform never needs to issue an explicit order. The creator internalizes the cues, adjusting preemptively to avoid the perceived consequences of noncompliance. The data is positioned as neutral truth, but in practice it is a filtered, curated representation of reality, one designed to serve the platform's commercial objectives. What is shown in the dashboard is a fraction of what the platform knows, and the selection of which insights to share is itself a form of behavioral engineering.

This asymmetry extends to the platform's use of the data it harvests. While the creator sees aggregated audience metrics, the platform's internal models are far more granular, incorporating psychographic profiles, purchasing behaviors, and cross-platform activity into a unified identity graph. These models allow the platform to monetize the creator's audience more effectively than the creator ever could, selling targeted ad space, cross-promoting other content, and even testing new monetization schemes using the creator's community as an unwitting focus group. The creator receives a fraction of the revenue generated, but their role as the primary attractor of that audience ensures that their data, like their content, remains an asset for the platform long after it ceases to benefit them directly.

The most sophisticated deployments of behavioral analytics blur the line between audience profiling and creator profiling. A platform can identify not only the content trends that are most likely to resonate with a given creator's audience, but also the emotional triggers and motivational levers that influence the creator themselves. If a creator is especially responsive to competitive ranking, they may be served more frequent leaderboard updates. If they react strongly to public recognition, their work might be spotlighted in curated features at moments calculated to sustain loyalty. The same datasets that optimize content delivery for viewers are used to optimize psychological containment for producers.

As this dual profiling matures, the platform gains an almost predictive power over the creator's trajectory. The algorithms are no longer simply reacting to what the creator has done; they are anticipating what the creator will do next, simulating multiple behavioral scenarios and nudging the one most beneficial to the platform's objectives. A creator who has shown responsiveness to trend adoption might suddenly see a stream of prompts encouraging them to use new features, join seasonal campaigns, or replicate formats that are currently monetizing well. These nudges appear as suggestions or opportunities, but they are, in effect, algorithmically generated instructions, carefully timed to maximize compliance without triggering resistance.

The carrot in this system is often an immediate and visible reward: a sudden boost in reach after following a prompt, an increase in engagement when using a recommended format, or a feature placement after joining a suggested campaign. The speed of the reward is critical. When a creator sees results within hours or days, they internalize the connection between compliance and success, making them more likely to respond to future nudges without hesitation. Over time, this reinforcement cycle reshapes the creator's instincts, replacing organic decision-making with algorithmically influenced reflexes.

The stick, by contrast, is the quiet withdrawal of these benefits when the creator deviates from the suggested path. A refusal to adopt a new feature might be followed by a subtle but measurable dip in visibility. Ignoring a campaign invitation could coincide with the platform surfacing competitor content more prominently in the same niche. These consequences are rarely presented as punishment; they are framed as the natural ebb and flow of audience attention. Yet the correlation is strong enough that creators learn to avoid the behaviors that trigger them.

The sophistication of behavioral analytics as both carrot and stick lies in its invisibility. Unlike overt policy enforcement, there is no single moment of conflict, no email warning of a violation. Instead, the creator experiences a gradual shaping of their incentives, a narrowing of the path that still appears to be freely chosen. They continue to feel autonomous even as their behavior converges toward the platform's ideal profile of a predictable, profitable contributor.

This shaping extends beyond content creation into the creator's business strategy. By selectively surfacing certain types of performance data, the platform can influence how a creator allocates their resources, what kinds of collaborations they pursue, which merchandise they produce, even where they focus their off-platform marketing. A dashboard that emphasizes short-term engagement over long-term retention, for example, encourages strategies that favor quick bursts of activity rather than sustained audience loyalty. The platform benefits from the resulting churn, as it can monetize spikes more effectively than slow, steady growth.

The endgame of such data harvesting is not simply to understand the creator, but to make the creator more understandable, more consistent, more legible, more easily integrated into the platform's revenue model. Every datapoint collected, every metric shared or withheld, moves the creator closer to a state where their output is not just compatible with the platform's systems, but dependent on them for validation, visibility, and survival. By the time this dependency is complete, the line between genuine creative choice and algorithmically engineered behavior is so thin as to be almost meaningless.

Why minorities and first-generation immigrants are prime targets

The targeting of minorities and first-generation immigrants by major platforms is neither incidental nor benign. It is the calculated exploitation of intersecting vulnerabilities, economic precarity, limited institutional backing, and systemic bias, that make these groups disproportionately susceptible to long-term capture. The process begins with the recognition that talent emerging from these demographics often exists outside the protective scaffolding of established industry networks. Without the insulation of legacy affiliations, legal teams, or well-capitalized support systems, such creators and innovators enter negotiations with platforms from a structurally disadvantaged position, even when their work demonstrates extraordinary potential.

For many first-generation immigrants, the pursuit of opportunity is bound up with the necessity of immediate income. The cost of relocation, coupled with the obligation to send remittances or support extended family, creates an urgency that platforms can readily leverage. Offers of accelerated monetization, guaranteed visibility, or upfront bonuses are framed as lifelines, fast tracks to stability in environments where traditional pathways to success are slower, riskier, or outright inaccessible. What is seldom disclosed is that these incentives are backloaded with restrictive terms: exclusivity clauses, derivative rights assignments, and algorithmic dependencies that secure the platform's long-term control over the creator's output. The offer is designed to feel too good to refuse precisely because refusal carries a real cost in the short term.

Minority creators face a parallel, though not identical, set of systemic constraints. Underrepresentation in mainstream media and technology sectors means that when a platform extends recognition, whether through curated features, partnership programs, or targeted outreach, it carries an amplified significance. The psychological weight of finally being "seen" by a powerful institution can lower resistance to contractual compromises that, under different circumstances, would warrant more scrutiny. This dynamic is further compounded by the scarcity of role models who have navigated similar opportunities without falling into dependency, leaving individuals to chart their course without a tested roadmap for maintaining autonomy.

Structural protection mechanisms, such as unions, guilds, or collective bargaining bodies, are either absent or underdeveloped in the digital sectors where these creators often operate. In industries like film or music, there are at least partial frameworks for protecting rights and negotiating minimum compensation. In the world of app development, online content creation, or niche cultural production, such protections are fragmentary at best. This absence leaves individual creators to negotiate directly with entities that command overwhelming leverage, both in terms of capital and in the ability to dictate market access.

Language and cultural barriers exacerbate the imbalance. Contractual language is frequently couched in jurisdiction-specific legal terminology that even native speakers may find opaque. For immigrants, particularly those still building fluency, the likelihood of misunderstanding key provisions, such as perpetual licensing rights or unilateral policy change clauses, is significantly higher. Platforms, while maintaining the appearance of accessibility, often provide little in the way of meaningful translation or legal guidance, leaving interpretation to the individual's own resources. This lack of clarity serves the platform's interests, allowing it to frame exploitative terms as "standard" or "industry norm" without effective challenge.

There is also a deeper economic calculus at play. Platforms understand that creators from marginalized or immigrant backgrounds often have fewer alternative revenue streams. Their dependence on the platform is not only professional but existential; losing access to monetization tools or algorithmic visibility can threaten their primary means of subsistence. This makes them less likely to engage in behavior the platform perceives as risky, speaking out against policy changes, experimenting with alternative distribution channels, or aligning with competitors. The platform's control is reinforced by the creator's own rational calculation that compliance, however uncomfortable, is safer than confrontation.

Psychologically, the targeting strategy also exploits narratives of upward mobility and inclusion. Marketing campaigns spotlight diverse success stories, presenting the platform as a democratizing force that "levels the playing field." While there are genuine instances of transformative opportunity, these are often exceptional cases used to obscure the structural reality: the vast majority of creators remain in a state of dependency, their autonomy constrained by the very systems that elevated them. The aspirational framing serves a dual purpose, it attracts new talent from similar backgrounds while reinforcing the loyalty of those already inside the system, who are led to believe that continued compliance is the path to becoming the next showcased example.

Once onboarded, the lack of structural protection means that any shift in algorithmic favor, monetization policy, or content guidelines can disproportionately harm these creators. A sudden change in payout thresholds or the introduction of new compliance requirements can collapse a precariously balanced livelihood overnight. Without legal recourse or a diversified income base, the creator is compelled to adapt instantly, often by doubling down on the very dependencies that caused the vulnerability in the first place. This reactive cycle benefits the platform, ensuring that the creator's focus remains on regaining lost ground within the ecosystem rather than exploring paths to independence.

In effect, minorities and first-generation immigrants become ideal "golden geese" in the platform's calculus: highly productive, culturally distinctive, and, because of systemic inequities, more malleable to long-term integration. Their innovations and cultural capital enrich the platform's brand, diversify its offerings, and open access to markets it might otherwise struggle to penetrate. Yet the economic rewards of this enrichment are disproportionately retained by the platform, while the creator's position remains as fragile as when they began. The vulnerability is not incidental, it is the very condition that makes them prime targets for capture.

The cycle becomes self-reinforcing as the platform uses its own analytics to identify and refine which vulnerabilities yield the highest long-term returns. Data gathered from onboarding programs, monetization behaviors, and content performance feeds into models that predict not just a creator's revenue potential, but also their likelihood of remaining loyal under increasingly restrictive conditions. Those from communities with limited institutional support, lower capital reserves, or fewer alternative distribution channels often score higher on retention metrics, not because they are more satisfied, but because the cost of leaving is too steep.

This predictive profiling allows platforms to fine-tune their offers. A creator from a well-connected background might receive a short-term collaboration or non-exclusive trial, while a first-generation immigrant with rapidly growing engagement might be offered a multi-year contract that front-loads incentives but back-loads obligations. The difference is strategic: the latter is more likely to perceive the offer as a once-in-a-lifetime chance and more willing to accept terms that quietly strip away their bargaining power over time.

For minority creators, especially those whose work draws from cultural heritage or niche community narratives, there is an additional layer of risk. Platforms are adept at commodifying cultural specificity, positioning it as a unique selling point in global marketing campaigns, while stripping it of context and ownership. Once integrated into the platform's ecosystem, the creative elements tied to a specific identity can be repackaged, algorithmically amplified, and even imitated by larger, more resource-rich producers. The original creator often receives minimal recognition or compensation beyond their initial contract, even as their cultural contributions drive significant engagement and revenue for the platform.

The absence of structural protection magnifies the harm when these inequities surface. Without collective representation or access to industry-standard legal advocacy, most affected creators must either accept the exploitation as the cost of visibility or attempt to fight it individually against entities with vastly superior resources. The imbalance is such that even public exposure of unfair practices rarely shifts the outcome; the platform can quietly recalibrate its promotional algorithms, replacing one voice with another while maintaining the same audience capture.

In this environment, trust becomes a double-edged sword. Many first-generation immigrants and minority creators enter partnerships with an assumption, shaped by cultural narratives about meritocracy, that demonstrated excellence will naturally lead to fair treatment. Platforms are skilled at reinforcing this belief through curated praise, incremental rewards, and selective showcasing of "success stories" that suggest the system is working as promised. By the time a creator recognizes the structural constraints, they are often deeply embedded in an identity as a "platform success," making departure feel not only economically risky but also like a personal repudiation of their own achievements.

This combination of economic precarity, cultural exploitation, and psychological conditioning explains why these demographics are disproportionately represented among long-term captive creators. It is not a matter of naivety or lack of ambition; rather, it is the result of targeted strategies that identify and exploit systemic vulnerabilities while concealing the mechanics of control behind the language of opportunity. In the platform's calculus, the very barriers that once excluded these voices from mainstream industries now make them the most strategically valuable assets, productive, loyal, and unlikely to defect.

Chapter 3, Tools of Enslavement

Policy as a Weapon

The wielding of policy as a weapon begins with the calculated ambiguity embedded in a platform's terms of service. These documents, often running to dozens of pages, are framed as protective instruments, legal frameworks to safeguard users and maintain a fair environment. Yet their true power lies in their elasticity. Language is deliberately crafted to be broad enough to encompass any scenario the platform may wish to act upon, now or in the future. Phrases such as "content that may harm the community," "activity that undermines the integrity of the service," or "behavior deemed inappropriate at our sole discretion" function as open-ended clauses, catch-alls that require no precise definition to justify suppression. The breadth of such wording ensures that enforcement can be highly selective, targeting individuals whose visibility or influence has become inconvenient, while leaving others unchallenged for similar conduct.

For the creator, the most dangerous aspect of this vagueness is that it transforms policy into a moving target. Today's tolerated behavior can be tomorrow's violation, not because it has inherently changed in nature, but because the platform's interpretation has shifted. This interpretive flexibility allows the platform to reframe enforcement actions as neutral application of the rules, even when those actions are strategically motivated. A creator advocating for more equitable revenue sharing may suddenly find their work flagged for "misleading claims." A competitor to a platform-owned product might be cited for "spammy promotion." These decisions can be cloaked in the authority of compliance, but their underlying purpose is often the preservation of commercial advantage.

The opacity of enforcement mechanisms compounds the problem. When a creator's content is removed, their account suspended, or their monetization disabled, the accompanying explanation is often little more than a restatement of the policy clause itself. Rarely are specific pieces of evidence disclosed, nor is there transparency about the decision-making process, whether it was the result of algorithmic detection, user reporting, or manual review. Appeals processes, where they exist, are typically routed through the same opaque system, with outcomes delivered in templated language that reveals nothing about the rationale. This lack of transparency serves a strategic function: it disorients the creator, leaving them unsure of which specific actions to avoid in the future, and more likely to self-censor broadly to stay within the platform's good graces.

This uncertainty creates a climate of anticipatory compliance, where creators modify their behavior not in response to explicit guidelines, but in deference to the platform's unstated preferences. The chilling effect is amplified by the visible consequences suffered by others in their network. Watching a peer lose their audience overnight due to a sudden policy enforcement sends a clear message: the safest course is to remain predictable, non-confrontational, and aligned with the platform's commercial narratives. In this way, vague terms of service do not merely regulate behavior, they shape it preemptively, exerting influence even in the absence of direct enforcement.

The selective application of vague policy also enables quiet retaliation against those who seek independence. A creator developing parallel distribution channels might suddenly find their account under "review," their reach throttled, or their payout delayed due to "verification issues." None of these actions need to be framed as punitive; each can be justified within the language of the terms. Yet the timing and pattern reveal the intent: to remind the creator that their operational stability depends on remaining in compliance, and that compliance is defined by the platform alone.

This strategic use of policy extends beyond suppression into the realm of competitive control. When a platform launches a new feature or product line, terms of service may be quietly updated to restrict behaviors that could divert attention or revenue away from it. These changes are often buried in "updates to our policies" notices that few users read in detail, yet they have the effect of narrowing the permissible scope of creator activity. By the time a creator realizes that their once-permissible monetization method or cross-platform promotion is now considered a violation, they may have already lost critical audience engagement or income.

Perhaps the most insidious function of vague terms is their ability to normalize overreach. Once a clause has been invoked a sufficient number of times, its application becomes accepted practice, regardless of its original intent. What begins as a safeguard against genuine harm can, over time, become a default justification for curating the platform's ecosystem according to its own strategic interests. The elasticity of language ensures that policy can always be adapted to emerging threats, whether those threats are to community well-being or to the platform's profit margins.

By maintaining the dual posture of protector and arbiter, the platform can suppress dissent, stifle competition, and enforce conformity without ever appearing overtly authoritarian. Each action is couched in the language of safety, trust, and compliance, making it difficult for creators to rally opposition without seeming to oppose those very values. The end result is a governance structure where the rules are both omnipresent and unknowable, where the boundaries of permissible behavior shift without warning, and where the creator's survival depends on reading signals that may or may not exist. In such an environment, policy ceases to be a set of rules to follow and becomes, instead, a weapon to wield, a flexible instrument of control that preserves the appearance of fairness while ensuring the primacy of the platform's interests.

Technical Control

Technical control operates as the silent enforcer of platform dominance, using the very infrastructure that enables creation and distribution to regulate the boundaries of a creator's autonomy. While contractual clauses and policy language provide the legal scaffolding for suppression, the decisive levers of enforcement are embedded deep within the technical systems themselves. Account suspensions, shadowbans, and feature throttling function not as isolated punishments but as integral components of a broader behavioral management framework, tools calibrated to maintain compliance, suppress dissent, and prevent the migration of value outside the platform's ecosystem.

The most overt of these tools is account suspension, an act that instantly severs the creator's connection to their audience, revenue streams, and accumulated content library. Publicly, suspensions are framed as necessary responses to clear violations of terms of service, but the internal decision-making process is rarely transparent. Automated systems scan for "violations" using pattern-matching algorithms, often flagging ambiguous content that human reviewers can interpret in multiple ways. This ambiguity allows enforcement to be highly selective: two creators can commit identical acts, but only one faces suspension, depending on their perceived alignment with the platform's strategic priorities. The suspended account becomes a cautionary tale, reinforcing the perception among other creators that the rules are both absolute and unknowable, and that the safest path is unquestioning adherence to the platform's preferences.

Shadowbanning is more insidious because it hides its own existence. The creator remains able to post, interact, and even see their own content in situ, but the visibility of that content to others is quietly reduced or eliminated. Engagement drops without warning or explanation, and the creator is left to question their relevance, quality, or audience loyalty. This uncertainty is part of the design. By withholding explicit acknowledgment of suppression, the platform avoids triggering formal appeals or public controversy. Instead, it induces a form of self-censorship as the creator experiments, often fruitlessly, to regain lost visibility. The hidden nature of shadowbans makes them uniquely effective: the creator continues to labor within the system, unaware that the system itself has decided to mute their voice.

Feature throttling operates at a more granular level, targeting specific capabilities rather than entire accounts. A live-streaming option might suddenly become unavailable, monetization tools might be restricted, or upload limits might be reduced. In many cases, these changes are explained as technical glitches, beta program closures, or policy "harmonizations." But the selective timing and application often reveal a different logic: throttling is deployed to nudge behavior, to steer creators toward desired patterns of content production or to disincentivize activities the platform finds unprofitable or competitive. A creator promoting off-platform links might find their clickable URL field disabled. One experimenting with formats that do not align with current monetization models might lose access to premium analytics or promotional boosts. Each adjustment is reversible, which makes it a flexible mechanism for exerting control without leaving a permanent mark on the creator's record.

What unites these mechanisms is their capacity to apply pressure without overt confrontation. A suspension is a blunt interruption, but shadowbans and throttling are gradual constrictions, technical equivalents of tightening a valve until the flow of attention, revenue, or functionality is just enough to keep the creator invested, but not enough to empower them to operate independently. This modulation can be finely tuned: exposure can be increased to encourage adoption of a new feature, then decreased to normalize the creator's dependency on that feature's revenue stream; throttling can be lifted for a period of compliance and reinstated after a deviation. In each case, the platform remains the unacknowledged architect of the creator's fortunes.

Over time, these controls condition creators to interpret stability as success. If their reach remains steady, they assume they are performing well; if it dips, they look inward for fault rather than recognizing the levers being pulled externally. This internalization of platform logic is the ultimate objective of technical control. The creator begins to preemptively align their behavior with what they believe the system rewards, avoiding risks, controversies, or experiments that might trigger invisible penalties. In this way, the architecture of technical control becomes self-reinforcing: the less it must be used overtly, the more effective it has become.

The power of these tools lies not only in their individual functions but in their orchestration. A creator facing a shadowban may simultaneously experience feature throttling that limits their ability to recover visibility, or find themselves under threat of suspension for "repeat violations" based on the content they have posted to counteract the suppression. This layering of constraints ensures that even the most resourceful creators are contained, their movements limited within a narrowing corridor of acceptable activity. And because these controls are embedded in the same systems that make creation possible, they cannot be bypassed without abandoning the platform entirely, a step that, for most creators, is economically and socially prohibitive.

Thus, technical control operates as both the infrastructure of empowerment and the machinery of subjugation. It grants the tools of creation while holding the levers that can silence, slow, or sever at will. Its genius lies in its invisibility: for the audience, the platform appears as a neutral stage; for the creator, the mechanisms of limitation are indistinguishable from the natural ebb and flow of digital life. In this duality, the platform secures its most enduring form of control, not by openly restricting the golden goose, but by adjusting the conditions of the cage so subtly that the goose believes it is still free to fly.

Financial Throttles

Financial throttles function as one of the most potent and least publicly visible instruments of platform control, operating at the intersection of technical infrastructure, policy discretion, and psychological leverage. Whereas algorithmic suppression and feature restrictions can diminish a creator's reach, financial throttling strikes directly at the lifeblood of their operation, cash flow. In the creator economy, where most participants operate without the cushion of large reserves or diversified revenue streams, the interruption or destabilization of income exerts an immediate and often overwhelming pressure to conform. It is a mechanism that ensures dependency not only by shaping creative choices but by conditioning survival itself on compliance with opaque and shifting rules.

The most blunt manifestation of this control is the delay in payouts. For creators who rely on platform earnings as their primary or sole source of income, the cadence of payments is as critical as the amount. A predictable disbursement schedule allows for planning, paying contractors, covering operating expenses, and meeting personal obligations. When a payout is withheld or postponed without clear explanation, it disrupts not just a budget but an entire chain of dependencies. Platforms often justify these delays under the language of "verification reviews," "fraud prevention," or "compliance checks." Such phrasing suggests a neutral, procedural safeguard, yet the triggers for these interventions are often neither transparent nor consistently applied. A spike in earnings, a surge of traffic from a particular region, or even an algorithmic misclassification can prompt a hold that lasts weeks or months. During this time, the creator is expected to continue producing content, effectively working on credit for the platform without guarantee of timely compensation.

The power of payout delays lies in their asymmetry. The platform continues to monetize the creator's work, generating interest or deploying the retained funds within its own financial systems, while the creator absorbs all the strain of disrupted cash flow. This arrangement is almost never framed as punitive; it is presented as a routine safeguard. But for the recipient, the impact is indistinguishable from a sanction, especially when the delay coincides with other forms of suppression or policy scrutiny. The unspoken message is clear: the stability of income is contingent on remaining in a state of constant eligibility, and eligibility is defined and redefined solely by the platform.

If payout delays are the slow constriction of a valve, sudden demonetization is the equivalent of shutting it off entirely. Here, the platform removes or disables the mechanisms by which a creator earns revenue, whether through advertising, subscriptions, tipping systems, or affiliate integrations, without necessarily suspending the account or restricting content posting. The effect is surgical: the creator's visibility may remain intact, giving the appearance of continued success, while their ability to extract value from that visibility vanishes. The justification is typically anchored in vague policy terms such as "advertiser suitability" or "brand safety." These phrases are elastic enough to encompass everything from the use of certain keywords to the discussion of sensitive topics, regardless of context or intent.

The swiftness with which demonetization can be enacted stands in stark contrast to the time it takes to reverse it. Creators frequently report waiting weeks or months for appeals to be processed, during which time their affected content remains revenue-dead. In some cases, entire back catalogs are swept into ineligibility due to a single policy update, retroactively erasing years of income potential. The economic effect is often catastrophic, but the psychological effect is equally potent: the knowledge that one's livelihood can be erased overnight without warning fosters a deep-seated aversion to risk. Even creators with strong principles or niche audiences learn to anticipate and avoid anything that could conceivably trigger the revenue kill switch, narrowing their creative range until it aligns with the platform's lowest-risk, highest-profit parameters.

In practice, financial throttles rarely operate in isolation. A payout delay can be paired with a temporary algorithmic downrank, ensuring that even if monetization is restored, the intervening dip in visibility has already eroded engagement metrics. Demonetization may coincide with a shift in content categorization, shunting a creator's work into less lucrative ad segments or subscription tiers. The layering of these tactics amplifies their effectiveness, making it clear to the creator that financial stability is inseparable from strategic alignment with the platform's operational priorities.

The most sophisticated deployments of financial throttling exploit the cyclical nature of creator earnings. Many platforms operate on models where revenue peaks during certain seasons, holidays for consumer goods, summer for travel content, election years for political commentary. By imposing restrictions during or just before these peak windows, platforms not only maximize pressure but also limit the creator's ability to recover lost ground. Missing a single seasonal revenue spike can represent a disproportionate share of annual earnings, and the fear of such a loss becomes a powerful deterrent against actions that might invite scrutiny.

Because financial throttles are rooted in administrative discretion and algorithmic opacity, they are difficult to contest. Appeals processes, when they exist, often return little more than a reaffirmation of the original decision, couched in the same generalized policy language that justified it in the first place. The lack of specificity means that even a successful appeal offers no clear guidance for avoiding future infractions, leaving the creator in a perpetual state of caution. This uncertainty is not a flaw in the system; it is a feature. It ensures that compliance is not just about following stated rules, but about intuiting and internalizing the platform's unstated preferences.

Over time, this financial modulation shapes the entire economic landscape of the platform's ecosystem. Creators who adapt quickly to its implicit demands, producing advertiser-friendly, engagement-optimized, controversy-free content, are rewarded with consistent monetization and even promotional boosts. Those who resist or miscalculate are marginalized not only in visibility but in earning potential, creating a two-tier system where the top layer reflects the platform's idealized output and the lower tier serves as a cautionary example. The result is an environment where economic viability is less a function of audience demand than of alignment with the platform's commercial architecture.

In its most extreme form, financial throttling can be used to drive acquisition. A creator whose income has been repeatedly disrupted may become more receptive to buyout offers, exclusivity contracts, or integration into platform-owned production programs. These arrangements promise stability, regular paychecks, guaranteed promotion, in exchange for surrendering control over content, branding, and distribution. What began as a pattern of inexplicable delays and demonetizations resolves into a coerced consolidation, with the creator effectively trading autonomy for economic security under the same entity that undermined it.

The genius of financial throttling as a control mechanism lies in its plausibility. Delays can be explained as security measures, demonetization as quality assurance, revenue fluctuations as market variance. To outside observers, and sometimes even to the creators themselves, these events can appear random or self-inflicted. The lack of visible cause-and-effect shields the platform from accusations of targeted suppression, even as patterns emerge that reveal its strategic use. It is the quiet precision of this tool, its ability to destabilize without direct confrontation, that makes it indispensable in the broader system of creator capture.

By turning the flow of income into a conditional privilege rather than an inherent right of participation, platforms ensure that every creator operates within a framework where economic survival depends on a continual, self-enforced calibration to the platform's evolving demands. In this way, financial throttles do more than shape individual behavior, they define the contours of the creative economy itself, determining not just who thrives, but what kinds of expression are economically sustainable in the long term.

Psychological Warfare

Psychological warfare within the platform ecosystem is not waged with blunt commands or overt threats, but through a carefully calibrated sequence of signals, reinforcements, and subtle manipulations that guide a creator's decisions without ever explicitly dictating them. It thrives on the appearance of choice, cultivating an environment in which compliance feels voluntary, even self-directed, while dissent is framed as irrational, risky, or self-sabotaging. The objective is not merely to secure short-term alignment but to reshape the creator's internal compass so that their sense of what is possible, desirable, and prudent falls naturally within the platform's strategic boundaries.

This conditioning begins with the constant presence of platform messaging, personalized dashboards, congratulatory notifications, reminders of milestones, and invitations to participate in curated programs. The tone is unfailingly positive, even when the underlying message is corrective. A decline in engagement is framed not as a symptom of algorithmic suppression, but as a prompt to "optimize posting times" or "try trending formats." This framing shifts responsibility for performance entirely onto the creator, erasing the platform's own role in distributing visibility. The creator is thus nudged to interpret every dip as a personal failure to adapt, reinforcing the belief that success lies in perfecting one's own output rather than questioning the distribution mechanisms that determine its reach.

Over time, the rhythm of these nudges becomes a background hum in the creator's working life. Notifications arrive with just enough regularity to keep the platform top-of-mind, yet with enough variation to feel organic rather than scripted. A congratulatory banner for reaching a certain follower count might appear on the same day as an email suggesting new monetization opportunities, subtly linking achievement to deeper integration into the platform's offerings. These messages rarely contain explicit instructions, yet they imply a roadmap: the path to greater success runs through the adoption of platform tools, adherence to recommended content strategies, and continued engagement with its community spaces.

The sophistication of this messaging lies in its capacity to mix praise with pressure, creating a dynamic in which the creator's self-esteem is partially outsourced to the platform's recognition systems. Public features, badges, or highlights in curated feeds serve as visible rewards, broadcasting the creator's status not just to them but to their peers and audience. The loss of such recognition, whether through the platform's deliberate choice or algorithmic rotation, registers as a social and professional setback, prompting the creator to seek reentry into the spotlight. This dynamic mirrors intermittent reinforcement schedules found in behavioral psychology, where the unpredictability of rewards fosters stronger and more persistent engagement than consistent reinforcement ever could.

As dependence deepens, the nudges grow more targeted. Platforms deploy data-driven segmentation to identify which creators are most responsive to certain types of prompts, competitive challenges, collaborative campaigns, or appeals to audience growth, and tailor their messaging accordingly. A creator who responds strongly to competition may receive invitations to ranking-based programs or leaderboards; one motivated by community might be offered the chance to "mentor" newer creators within a closed group. While these opportunities appear to serve the creator's interests, they also function as anchors, tethering them to the platform's internal economy of prestige and influence.

The psychological leverage is most visible in how platforms frame departures from their preferred trajectory. Creators exploring off-platform ventures may suddenly receive a spike in engagement on platform-exclusive features, timed to coincide with these experiments. The message is implicit but unmistakable: attention and opportunity reside here, not elsewhere. In more coercive iterations, the platform's messaging will stress the "risks" of dilution, warning that splitting focus between multiple channels may confuse audiences or weaken the creator's brand. This narrative is reinforced by selective amplification, ensuring that content aligned with exclusivity receives the most prominent placement. The creator is thus guided to view independence not as diversification but as fragmentation, a potential loss rather than a gain.

What makes these tactics so effective is that they operate within the creator's own cognitive and emotional frameworks. The platform positions itself as coach, collaborator, and confidant, using language and timing that align with the creator's aspirations and anxieties. When the creator begins to internalize this guidance as self-generated insight, believing they are making purely autonomous choices when they are, in fact, following a conditioned path, the psychological capture is complete. The boundaries between independent judgment and externally influenced decision-making blur to the point where they are indistinguishable.

This is not achieved through a single interaction, but through a sustained interplay of micro-signals over months or years. Metrics dashboards that emphasize certain KPIs over others can slowly shift a creator's definition of success. Notifications that highlight high-performing content subtly redefine what is "worth" producing. Even the absence of messaging, periods of silence from the platform, can be leveraged to create anticipation, making the next prompt or opportunity appear more valuable than it objectively is. Each of these elements, taken alone, might seem benign or even helpful; taken together, they form a system of psychological governance as robust as any technical or financial control.

In the end, the erosion of independence is not experienced as a hostile takeover, but as a gradual narrowing of the imagination. The creator's vision of what they can do, who they can reach, and how they can define success becomes so entwined with the platform's infrastructure that alternatives feel implausible, impractical, or unsafe. It is in this state, when the cage has been internalized as the horizon, that psychological warfare achieves its most enduring victory.

Case study deep-dive

When she began, her online presence was little more than a personal experiment, a modest channel on a mid-tier video platform where she posted short, skill-based tutorials filmed in the corner of her apartment. The early growth was slow but steady, each new subscriber hard-won through hours of engagement in comment threads and participation in community forums. The lack of corporate attention afforded her an unusual degree of creative freedom. She could experiment without fear of audience alienation because the audience, at that point, was intimate enough to follow her through failed experiments as readily as through successes. This autonomy was the soil in which her voice matured. It also meant that when the first platform extended its hand, the offer felt like validation rather than the opening move of a long campaign to integrate her into a controlled economy.

The invitation came in the form of an email from a "Creator Partnerships" liaison, congratulating her on surpassing an internal engagement threshold. The offer promised elevated visibility, access to advanced analytics, and priority inclusion in their curated "learn" category. The terms were not exclusive in the legal sense, but the onboarding process subtly reoriented her workflow. The new analytics dashboard emphasized specific metrics, average view duration, click-through rates on suggested videos, that, while useful, also began to frame her understanding of success in platform-centric terms. The liaison encouraged her to tailor future content to maximize these numbers, noting how doing so could trigger "algorithmic momentum" and lead to exponential growth. Without realizing it, she was already aligning her creative compass with an external magnet.

Within months, she experienced her first significant spike in visibility. The curated placement brought in a wave of new viewers, but it also created a dependency she could not yet see. When that placement rotated out, her traffic dropped sharply, and the liaison was quick to suggest participation in a beta program for interactive overlays, a feature that would both increase watch time and integrate her content more deeply into the platform's proprietary technology. She agreed, seeing it as the natural next step, unaware that the format's technical demands would make it cumbersome to export her content to other ecosystems.

Her growing reputation caught the attention of a larger, global platform whose audience dwarfed the one she currently served. Their outreach was more formal: a multi-year partnership proposal with guaranteed monthly payments in exchange for priority distribution rights to her new releases. The upfront stability was appealing; it would allow her to quit her part-time job and focus entirely on content creation. But the contract contained provisions that mirrored what she'd already informally conceded on the first platform, concentration on certain performance metrics, participation in quarterly promotional campaigns, and early adoption of platform-led features. Migration was not framed as a severance from the first platform, but as an expansion. In practice, however, the time and resources required to meet the demands of the new arrangement meant her output on the original channel slowed to a trickle, leading to its gradual decline.

The second platform's systems were more sophisticated in their behavioral conditioning. Automated prompts suggested optimal posting times down to the minute, nudged her toward trending topics with high advertiser demand, and delivered weekly performance summaries that celebrated adherence to these cues. The financial terms meant she could not afford to ignore these nudges; missing growth targets could trigger clawbacks or reductions in the guaranteed minimum payment. The sense of independence she had once taken for granted was now tethered to compliance with an invisible but constant rhythm.

It was during her second year on the global platform that she first encountered financial throttling. A sudden "account review" placed her monthly payout on hold, citing the need to verify certain advertiser safety requirements. No specific violation was identified, and the liaison handling her account was vague about the timeline for resolution. For six weeks, she continued to post at the expected cadence, effectively producing content for free while the platform continued to monetize it. When the hold was finally lifted, the payout was released in full, but the experience left a residue of anxiety. She understood now, viscerally, that her income could be interrupted without warning, and that her only protection lay in avoiding anything that might draw scrutiny.

Around this time, she opened an account on a third platform, a short-form video app whose explosive growth made it a coveted source of audience expansion. Initially, her intent was to use it as a funnel, posting teasers that would direct viewers to her main channel. But the app's recommendation algorithm proved intoxicating. A single clip could reach hundreds of thousands overnight, dwarfing the numbers she achieved elsewhere. The platform capitalized on this, inviting her into its "Rising Stars" program, which offered bonuses for meeting weekly viewership goals and experimenting with trending audio tracks. The incentives pulled her deeper, and soon she was producing original content exclusively for this app to maintain momentum. The other platforms noticed.

Subtle consequences followed. On the global platform, her homepage features became less frequent. Promotional support was offered only if she agreed to produce exclusives that would not appear on the short-form app. The smaller, original platform adjusted its algorithm so her infrequent uploads there reached only a fraction of her remaining audience. No one told her she could not be multi-platform, but the coordinated withdrawal of support made the costs clear.

When a content moderation sweep on the short-form app flagged several of her videos for "community guideline violations," her bonus eligibility was suspended. Again, no specific infractions were detailed beyond broad policy categories. The flagged videos were educational and non-controversial by any reasonable standard, but they shared a common trait: each contained a verbal mention or visual prompt directing viewers to her other channels. Whether by design or coincidence, the effect was the same, the most efficient pathways for her to guide audience migration were now obstructed.

By her fifth year in the digital economy, she was fully enmeshed in three ecosystems, each applying its own blend of technical controls, policy levers, financial modulation, and psychological nudges. The original platform retained archival value for her loyal base but generated little income. The global platform provided stability but demanded conformity to a narrow, advertiser-friendly mold. The short-form app offered explosive reach but at the cost of volatile moderation policies and an unpredictable revenue model. Each platform monitored her activity on the others, rewarding exclusivity and punishing divergence through mechanisms subtle enough to maintain plausible deniability.

Attempts to reclaim independence were met with swift, if understated, countermeasures. A crowdfunding campaign launched on an external site to fund an independent content hub saw its promotional posts suppressed across two of her three primary platforms. Traffic from her largest audience source to the campaign page was inexplicably low compared to historical click-through rates, suggesting algorithmic downranking. Direct outreach to her subscribers via email newsletters faced new hurdles as platforms began restricting the visibility of posts containing certain types of external links.

The cumulative effect was not simply professional fatigue but a profound recalibration of her ambitions. In the beginning, success had meant growth on her own terms, the freedom to experiment, and the ability to build a sustainable livelihood around her creative vision. Now, success was defined by stability within the conditions set by her host platforms, a steady income, consistent visibility, and avoidance of actions that might trigger punitive responses. The parameters of her career had been redrawn so gradually, through so many interlocking mechanisms, that she could no longer pinpoint the moment when her independence had been replaced by managed containment.

Her story illustrates the compounded effect of cross-platform capture. The tactics described in isolation, financial throttles, policy ambiguity, technical suppression, and psychological conditioning, are potent on their own. But when deployed across multiple ecosystems, they create a web from which escape is exponentially more difficult. Each platform's enforcement reinforces the others, not necessarily through formal collusion but through shared incentives and mirrored strategies. A decision made on one platform echoes across the rest, amplifying the cost of resistance and minimizing the viability of alternative paths.

In private moments, she considered walking away entirely, imagining a return to the modest, self-directed operation she had once maintained. Yet the infrastructure that sustained her work, audience relationships, monetization systems, and promotional pipelines, was now so thoroughly embedded in platform architecture that dismantling it felt akin to dismantling her livelihood. The very tools that had accelerated her growth had become the walls of her confinement, and the freedom she once sought was now something she would have to rebuild from the ground up, outside the gates of the systems that had, for years, defined her professional identity.

Chapter 4, The Economics of Captivity

Value Extraction Mechanics

The mechanics of value extraction in the platform economy are hidden in plain sight, embedded in the very systems that promise empowerment. At the surface level, creators are offered a clear and apparently equitable proposition: produce content, attract an audience, and earn revenue proportionate to the engagement and loyalty you generate. Yet the proportionality is an illusion. The reality is that the intellectual property you create and the network effects it triggers generate far more value for the host platform than is ever returned to you. This disparity is not a byproduct of inefficiency; it is the foundational design of the system.

Every upload, stream, or post is the visible tip of a much larger economic iceberg. The creative work itself, be it a video, an image, a software application, or a piece of writing, functions as the magnet that draws user attention. Once that attention is captured, the platform monetizes it through multiple revenue streams that extend far beyond the monetization pathways made available to the creator. Advertising is the most obvious, but even here the split is rarely close to equal. Platforms sell inventory around, before, after, and sometimes directly within the creator's content, leveraging premium ad rates when the demographic targeting aligns with advertiser demand. The creator's share is calculated not from the full value of these placements but from a reduced subset, adjusted by opaque formulas, subject to deductions for "invalid traffic," "brand safety concerns," or "platform service fees."

Beyond direct ad sales, platforms monetize the behavioral data generated by the audience's interaction with the creator's work. Every click, pause, replay, comment, and share feeds into a data model that enhances the platform's ability to predict and influence user behavior. This data is not sold in raw form, something that would draw regulatory scrutiny, but it is packaged into targeting capabilities that advertisers pay a premium to access. The creator, whose content catalyzed the interaction that produced the data, receives nothing from this secondary value stream. The fact that a single high-performing piece of content can materially improve a platform's recommendation algorithms across millions of unrelated accounts underscores the scope of this uncredited contribution.

Then there is the network effect. A creator who draws in a new user does more than deliver one set of eyeballs to one piece of content. That user becomes part of the platform's ecosystem, generating value for years through interactions with other creators, purchases in affiliated marketplaces, or subscriptions to premium services. The original creator is rarely, if ever, compensated for the lifetime value of that acquisition, even though traditional affiliate marketing models in other industries routinely recognize and reward customer acquisition. On platforms, the referral is absorbed into the general traffic flow, the attribution link is erased, and the ongoing value accrues entirely to the host.

The gap between generated value and received compensation widens further when we consider derivative monetization. A creator's content can be used to promote platform-wide initiatives, train machine learning models, or populate curated collections that enhance the platform's brand identity. In each of these cases, the content serves purposes beyond its original context, often without additional consent or compensation. The platform's terms of service, accepted with a single click during account creation, almost always include licensing provisions broad enough to permit such uses indefinitely. The economic benefit of these uses flows exclusively to the platform, while the creator's compensation remains tied to the performance of the original post in its original setting.

Merchandising and sponsorships represent another layer of disparity. Platforms increasingly act as intermediaries in connecting creators with brand deals, but they extract fees and control the terms of these transactions in ways that favor their own interests. Even when the platform is not directly involved, the visibility it grants or withholds can determine whether such opportunities arise at all. A creator who falls out of algorithmic favor may see inbound offers dry up, regardless of audience loyalty, because brands prefer to work with those the platform is actively promoting. In this way, the platform's gatekeeping over exposure indirectly controls access to off-platform revenue, further consolidating its influence over the creator's total earning potential.

The asymmetry becomes most apparent when we examine the economics of scale. A single viral video might generate tens or even hundreds of thousands of dollars in ad revenue for a platform, not counting the additional value from user retention, cross-promotion, and data harvesting. The creator might receive a fraction of this sum, sometimes less than ten percent, despite being the sole originator of the asset that made the windfall possible. At scale, across thousands of creators and millions of uploads, the aggregate difference between what is generated and what is paid out becomes the core of the platform's profitability. This is why platforms can afford to offer free hosting, unlimited bandwidth, and global distribution: the surplus value extracted from creators more than covers the operational costs.

For the creator, this gap is often obscured by the opacity of reporting. Revenue dashboards provide granular detail on what the creator has earned, but little to no insight into the total value their content has generated for the platform. The metrics that could illuminate this disparity, total ad impressions sold against the content, effective CPMs achieved in premium placements, data monetization multipliers, are closely guarded trade secrets. By controlling the informational environment, the platform ensures that creators calibrate their expectations to the numbers they see, rather than to the full value their work produces.

This structure also creates a psychological alignment that works in the platform's favor. Because creators see their own revenue fluctuate in response to audience engagement, they naturally attribute changes to their own performance, not to shifts in the platform's monetization strategy. If a payout is lower one month, they may assume they failed to maintain interest, not realizing that the platform has adjusted ad load, pricing, or revenue share percentages to meet internal targets. The invisibility of these levers means that even experienced creators often underestimate how much of their earning potential is controlled externally.

When platforms do adjust payout structures, the changes are typically framed as necessary optimizations to "support the community" or "improve sustainability." Rarely are they presented as margin-protection measures, though that is often their primary function. These adjustments can include lowering revenue shares, tightening eligibility criteria, or reclassifying certain types of content as less valuable to advertisers. Each change reduces the proportion of value returned to creators, but because the total pool is so opaque, the cumulative effect can be difficult to measure. Over time, however, the percentage of platform-wide revenue allocated to creators tends to decline, even as total platform revenue grows.

The imbalance is not solely a matter of greed; it is also structural. Platforms operate in a winner-takes-most environment, where network effects and scale advantages make it more efficient to invest in infrastructure and product development than to increase payouts proportionately. From the platform's perspective, the creator's content is one of many interchangeable inputs into a system designed to maximize user time-on-platform. The marginal utility of any single creator's contribution is high when they are small and growing, but diminishes as the platform's content library expands. This allows the platform to keep payouts low without risking a significant drop in user engagement, because there is always another creator ready to fill the gap.

For the creator, the long-term consequence of this system is a persistent undervaluation of their intellectual property. Unlike traditional licensing arrangements in which rights holders negotiate terms based on the projected value of their work, creators on platforms typically grant broad usage rights in exchange for a share of revenue that is both limited in scope and variable in rate. The licensing is perpetual, the rights expansive, and the compensation contingent, not on the intrinsic worth of the work, but on its performance within the narrow confines of the platform's monetization channels.

This disconnect between generated value and received compensation is the economic heart of the captivity described throughout this book. It is not enough to understand the creative, technical, and psychological mechanisms of control; one must also see the financial architecture that makes them viable. Platforms can afford to suppress reach, throttle features, delay payouts, and condition behavior precisely because the surplus they extract is so large. That surplus is the buffer that absorbs the cost of experimentation, the risk of attrition, and the expense of maintaining an ever-expanding technological and legal apparatus to manage the creator population.

From a macroeconomic perspective, this model resembles other extractive industries, where those who control the means of distribution capture the majority of value, leaving originators with a fraction of the wealth their labor produces. In resource economies, this dynamic has been criticized for fostering dependency, discouraging innovation, and perpetuating inequality. In the digital creator economy, it has been normalized under the rhetoric of democratization and access. The language of empowerment masks a familiar equation: control over infrastructure begets control over value distribution, and control over value distribution ensures the perpetuation of the underlying power imbalance.

The Closed Loop

The closed loop begins with an almost imperceptible narrowing of choice. At first, the creator is free to select the tools, platforms, and channels through which they conceive, produce, and deliver their work. The market appears open, the infrastructure interoperable, the field of distribution wide and varied. Yet the architecture of the modern platform economy is designed to incrementally remove viable alternatives until production and distribution occur exclusively within a single, controlled ecosystem. This enclosure is neither accidental nor collateral, it is the strategic objective. By ensuring that every stage of the creative cycle, from ideation to monetization, is dependent on proprietary systems, the platform transforms itself from a service provider into an indispensable intermediary, one that sits between the creator and their audience at every point of contact.

The mechanisms of enclosure often masquerade as convenience. Proprietary editing suites, integrated analytics, and in-app monetization tools are offered as value-added services, ostensibly to save time and consolidate workflow. In reality, they bind the creator's process to formats, file structures, and feature sets that are not easily portable to competing environments. A video edited within a platform's native suite may be encoded with metadata optimized for that platform's algorithm, giving it a performance advantage that cannot be replicated elsewhere. The creator, eager for the best possible outcome, adapts their production habits accordingly, unaware that each adjustment deepens their technical dependency.

The distribution side of the loop is fortified through control over audience access. Platforms position themselves as the singular gateway to reach large, engaged communities. While in theory nothing prevents a creator from publishing on multiple channels, the algorithmic systems that determine visibility often favor exclusivity, rewarding content that is native to the platform and penalizing work that is shared externally. This is done subtly, through ranking adjustments, recommendation biases, and discoverability filters, so that the creator experiences the decline in reach as an organic shift in audience interest rather than as a deliberate suppression of cross-platform promotion. Over time, the path of least resistance, publishing and distributing entirely within the platform, becomes the only path that offers sustainable engagement.

Monetization is the final and most decisive lock in the loop. By hosting the means of payment within the ecosystem, whether through ad revenue sharing, subscription features, or direct tipping systems, the platform ensures that economic activity is both generated and captured internally. External monetization methods, such as off-platform merchandise sales or third-party sponsorships, are discouraged through restrictive linking policies, reduced promotional support, or direct competition from platform-managed alternatives. The creator learns that revenue potential is highest when it flows through the platform's own channels, not because those channels are inherently superior, but because the platform reserves its best audience targeting and promotional capabilities for transactions it controls.

Once the loop is closed, the creator operates in a fully contained environment. Their tools are platform-owned, their audience exists primarily within the platform's database, and their income is mediated by the platform's payment systems. This total integration grants the platform unparalleled leverage. Any dissatisfaction, over revenue splits, policy changes, or creative restrictions, is tempered by the reality that leaving would mean forfeiting not only distribution but also production infrastructure and the accumulated goodwill of an audience whose access is controlled by the very entity the creator might wish to escape.

The loop is self-reinforcing because it is designed to align with the creator's short-term incentives while eroding their long-term autonomy. Each stage of enclosure offers tangible, immediate benefits: easier workflow, larger audiences, more streamlined monetization. The costs, technical dependency, audience captivity, revenue gatekeeping, are deferred and diffused, becoming apparent only when the creator attempts to act outside the prescribed boundaries. By that time, the friction of exit has been engineered to be prohibitively high, both economically and operationally.

At scale, the closed loop does more than constrain individual creators; it reshapes the market itself. Competitors are starved of content because creators are incentivized to remain exclusive. Alternative distribution channels struggle to gain traction without access to the audiences already locked inside dominant ecosystems. Innovation in production tools is stifled because the most visible creators are developing content tailored to a single set of proprietary standards. The platform, having secured both supply and demand, can dictate terms to advertisers, sponsors, and even end users, consolidating its position as the indispensable intermediary in the entire value chain.

This consolidation has broader cultural consequences. When production and distribution are contained within a single ecosystem, the diversity of creative output is filtered through that ecosystem's commercial logic. Formats that maximize ad inventory or subscription retention are prioritized; those that do not are sidelined, regardless of their cultural or artistic merit. The feedback loop between creator incentives and platform priorities produces a homogenization of style, pacing, and subject matter, narrowing the expressive range of the medium. Audiences, conditioned by the platform's recommendation systems, come to expect and prefer content that conforms to these norms, further entrenching the cycle.

The closed loop is not static; it evolves with the platform's strategic needs. As new revenue models emerge, such as e-commerce integration, live event streaming, or NFT marketplaces, they are folded into the ecosystem, expanding the scope of activities that must occur within the controlled environment. Each addition increases the opportunity cost of leaving, as creators who invest in these new features find their audiences and revenues increasingly entangled with platform-specific mechanisms. Even when external markets exist for these activities, the platform leverages its scale and data advantage to make its internal options more attractive, if not outright necessary.

What makes the closed loop particularly resilient is its invisibility to those operating within it. Because the enclosure happens incrementally, each new layer of integration feels like a natural evolution rather than a strategic constraint. The creator, focused on day-to-day production and audience engagement, may not perceive the gradual loss of optionality until they face a decision that requires acting outside the system. By then, the loop has achieved its purpose: not merely to contain the creator, but to make the idea of operating elsewhere seem impractical, unprofitable, or both.

In this way, the closed loop becomes the ultimate instrument of capture. It does not rely on force or explicit prohibition. Instead, it shapes the environment so completely that the creator's most rational choice is to remain inside, producing and distributing in ways that maximize the platform's value extraction while minimizing their own ability to exist independently. The platform's control is total, yet it is exercised without visible coercion, a seamless, self-sustaining circuit in which every path leads back to the same gate.

Lost Opportunity Costs

Lost opportunity cost in the platform economy is rarely presented as a measurable figure, yet it defines the true scale of the imbalance between what you create and what you are allowed to capture. It is the shadow value, the income, reach, and long-term stability that would have been yours had the conditions been different, that remains unrecorded in any payout dashboard or performance report. This hidden ledger is not simply a matter of the revenue share you forfeit; it encompasses every dimension of value withheld through structural design, algorithmic manipulation, policy constraints, and market distortion. Calculating it is not an exercise in abstract grievance, it is the act of restoring visibility to what has been systematically erased from your economic reality.

The most immediate layer of lost opportunity is the difference between the gross value your work generates and the net amount you receive. This is the surface-level disparity visible in any revenue split arrangement. A video that draws in millions of views may be credited to you with a payout calculated from a small fraction of the ad placements it facilitated. The platform's retention of the majority share is often justified by hosting, distribution, and marketing costs, but these figures are never disclosed in proportion to the revenue retained. Without transparent accounting, the true profit margin on your work remains invisible, yet it is the foundation on which the platform's own expansion is built. That difference, the unremitted value, is the most obvious and easiest to imagine reclaiming in a different structural arrangement.

But the larger and more consequential component of lost opportunity cost is not in the share withheld from what is monetized; it is in the monetization that never happens at all because of choices the platform makes on your behalf. Algorithmic suppression can mean that content with genuine audience demand is never shown to those who would engage with it. The potential revenue from these unseen interactions does not appear as a loss in your reports, it vanishes before it ever reaches the ledger. Similarly, restrictions on external linking or cross-platform promotion can sever potential streams of off-platform income before they are even established. When you are prevented from directing an audience to your own storefront, your own mailing list, or your own subscription service, the compounded loss over time is far greater than any single month's ad revenue shortfall.

These costs are magnified when we factor in the lifetime value of missed audience connections. A user who is never shown your work due to throttled visibility is not simply a lost view; they are a lost relationship that could have supported multiple revenue streams over months or years. Platforms understand this principle well when calculating their own customer acquisition costs, yet creators are rarely encouraged, let alone allowed, to think about their audiences in similar terms. The deliberate prevention of such thinking is itself part of the control mechanism, ensuring that you focus on optimizing content for the platform's short-term metrics rather than on cultivating direct, enduring connections that could sustain you independently.

Another layer of loss occurs when your creative range is narrowed to fit the platform's commercial priorities. If you refrain from producing certain types of work because they are deemed "brand unsafe" or less likely to be promoted, you forgo not only the immediate revenue from that work but also the potential audience segments and future opportunities that might have emerged from it. This is a form of opportunity cost that is difficult to measure precisely, but its impact is profound. Over time, the suppression of less "profitable" creative choices shapes not only your income but your public identity, narrowing the market perception of what you can do and thereby reducing the range of commissions, collaborations, and licensing deals you are offered.

Financial throttles, such as delayed payouts or sudden demonetization, introduce yet another form of hidden cost. The delay in receiving funds can force you to postpone investments in equipment, marketing, or personnel that would have accelerated growth. The compound effect of such delays, measured not in days or weeks but in the months or years of momentum they disrupt, represents a significant subtraction from your long-term earnings. This is not a neutral deferral; it is a disruption to your ability to reinvest in your own enterprise at the optimal moments. The cost is both financial and strategic, as opportunities to capitalize on timely trends or market conditions pass by while your capital is locked in the platform's control.

There is also the strategic displacement that occurs when platforms use your work to draw audiences toward their own proprietary products, services, or favored partners. If your content appears alongside promotions for goods and services from which you receive no share, the audience's purchasing capacity is being harvested in ways that benefit others at your expense. The lost opportunity here is not only the missed direct sale but the erosion of the audience's spending potential for your future offers. Each conversion that could have been yours but is diverted toward the platform's preferred outcomes is a silent subtraction from your long-term earning horizon.

When platforms monopolize the distribution channel, they also dictate the terms of audience discovery. The architecture of search and recommendation determines who finds you, how often, and under what conditions. If these mechanisms prioritize content from entities with larger budgets, corporate partnerships, or other strategic advantages, your discoverability is reduced not by the market's natural preference but by the platform's engineered priorities. Every audience member who ends up in a competitor's funnel instead of yours is an incremental opportunity cost, a lost subscriber, supporter, or customer who may never be recovered.

The network effects of this dynamic compound the losses. High-visibility creators attract more opportunities simply because they are perceived as successful. This perception is often shaped by the very algorithms that control visibility, creating a self-reinforcing cycle in which those favored by the system become magnets for sponsorships, media coverage, and collaborations, while those deprioritized by the system see such offers decline or vanish entirely. The lost opportunity cost here is not merely the direct revenue from one missed collaboration; it is the cascade of subsequent opportunities that never materialize because you were not positioned in the right light at the right time.

Calculating these costs in precise monetary terms is difficult because they are designed to be invisible, yet they can be modeled conceptually. One might begin by estimating the total addressable audience for your work in a neutral distribution environment, then applying realistic engagement and conversion rates to that number. The difference between this model's projections and your actual reported outcomes is the gap created by the platform's control mechanisms. Even if such calculations are approximate, they illuminate the magnitude of what is withheld, not through theft in the conventional sense, but through the structuring of the environment to ensure that only a portion of the potential value ever reaches you.

The psychological effect of these hidden costs is itself a form of loss. When you operate in an environment where the full scale of your contribution is obscured, your benchmarks for success are set artificially low. You learn to celebrate incremental improvements within the confines of a system that is, by design, preventing you from accessing the majority of your potential market and revenue. This reframing of expectations is not accidental; it is a way of aligning your ambitions with the platform's interest in maintaining a stable, compliant creator base. By lowering your sense of what is achievable, the system reduces the likelihood that you will push for structural changes or seek independence.

The ultimate lost opportunity cost is not any single missed payment, collaboration, or sale. It is the forfeiture of a future in which your creative output is valued on terms that you define, in markets you control, with audiences you own. Every structural barrier, every policy constraint, every algorithmic tweak that limits your reach is a subtraction from that future. The tragedy is that you may never see its full shape; the possibilities you are denied do not announce themselves. They pass in silence, unmarked except in the quiet narrowing of your trajectory. The cost is measured not only in money but in the scale of the life and career you might have built had the system been aligned with your growth rather than its own containment.

Market Distortion

Market distortion under the regime of captive intellectual property is not a mere byproduct of the platform economy, it is one of its primary functions. When creative work is systematically enclosed within proprietary ecosystems and subjected to the rules, incentives, and limitations of those ecosystems, the broader industry ceases to operate according to the organic forces of supply, demand, and merit. Instead, the flow of capital, attention, and opportunity is redirected by mechanisms that serve the controlling entity's strategic priorities, often at the expense of the health, diversity, and sustainability of the market itself. What emerges is an environment where the value of work is determined less by its intrinsic merit or audience resonance and more by its compatibility with the commercial imperatives of a small number of gatekeepers.

The first and most visible form of distortion is the artificial inflation of content that aligns with the platform's short-term objectives, regardless of whether it reflects the broader tastes of the market. Recommendation algorithms, promotional placements, and curated showcases act as amplification systems that can elevate particular styles, topics, or formats far beyond the level they would achieve in a neutral distribution landscape. This does not merely shift audience attention temporarily, it recalibrates audience expectations, reshaping what is considered standard or desirable within a given genre. Over time, the artificially promoted aesthetic becomes the de facto industry norm, forcing independent producers to mimic it in order to remain competitive, even if doing so erodes their originality or undermines the integrity of their work.

The second form of distortion occurs through the systematic underexposure or devaluation of work that does not fit the platform's commercial model. Content that is slower to consume, harder to monetize, or less likely to attract certain categories of advertisers is algorithmically sidelined, regardless of its cultural significance or long-term audience appeal. In a neutral market, such works might flourish within dedicated niches, supported by loyal audiences willing to pay for them directly. Under a captive system, those niches are starved of discoverability, and their economic viability declines to the point where fewer creators can sustain themselves within them. The result is a narrowing of the creative landscape, with entire subfields atrophying because they cannot meet the distribution criteria of the dominant gatekeepers.

This narrowing of diversity has a feedback effect on the broader industry. As fewer high-quality works emerge from suppressed niches, the perception grows, among audiences, investors, and even aspiring creators, that these forms are inherently less valuable. This perception is then used to justify further suppression, creating a self-reinforcing cycle in which the absence of representation is mistaken for the absence of demand. In truth, the demand has not vanished; it has been systematically displaced by the structural preferences of the platform economy. The lost works, the displaced audiences, and the diminished creative ecosystems are all casualties of a market that no longer reflects the organic interplay between producer and consumer.

Financial arrangements within captive ecosystems further distort market norms by redefining the baseline for acceptable compensation. Revenue-sharing models, which allocate only a small fraction of gross value to the originator of the work, set a precedent that begins to seep into negotiations outside the platform environment. Clients, sponsors, and distributors accustomed to the low payout rates of dominant platforms may resist paying higher rates in independent transactions, even when the economics of those transactions justify it. Over time, the suppressed rates of the captive ecosystem become the unofficial benchmark for the industry at large, making it harder for creators to secure fair compensation in any context.

This wage compression is mirrored by an inflation of the platform's own value within the transaction chain. Because the gatekeeper controls both distribution and audience data, it can position itself as the most indispensable player in the market, claiming the lion's share of revenue even when its actual contribution to production is minimal. Advertisers, brands, and other intermediaries adjust their budgets accordingly, directing more spending toward platform-controlled inventory and less toward independent deals. The cumulative effect is a reallocation of resources away from the originators of intellectual property and toward the entities that control its circulation, further consolidating the imbalance of power.

The distortion extends into the domain of audience metrics, where the controlling entities define the standards by which success is measured. Engagement rates, view counts, and follower growth are all calculated within proprietary systems, which can be tuned to favor certain behaviors or formats. These metrics, in turn, become the currency of credibility within the industry. A creator's ability to secure funding, collaborations, or media coverage may hinge on performance indicators that are themselves products of the platform's internal logic, rather than impartial reflections of audience interest. This creates a situation in which the very definition of success is a function of one company's algorithmic preferences, and where the work required to achieve that "success" is indistinguishable from the work of conforming to the system's commercial agenda.

One of the most insidious effects of this distortion is its impact on innovation. In a healthy market, creators and producers can experiment with new formats, business models, and audience relationships, knowing that success will be determined by the resonance of their ideas in the open field. In a captive market, the cost of deviation is disproportionately high. Innovation that does not immediately align with platform priorities is likely to be buried in distribution, depriving it of the oxygen needed to prove its viability. Over time, this suppresses the appetite for risk-taking, both among creators and among the investors or partners who might support them. The result is an industry skewed toward iterative refinements of existing formulas, optimized for the quirks of the current gatekeeper's system rather than for the evolving needs and interests of the audience.

Even industries adjacent to the platform's primary focus are reshaped by the gravitational pull of its norms. Music, publishing, film, education, and live events all adapt their formats, release schedules, and marketing strategies to align with the patterns established in the captive digital space. This convergence toward platform-optimized norms erodes the distinctions between mediums, flattening the cultural landscape and reducing the range of expressive possibilities. In effect, the tail begins to wag the dog: instead of distribution adapting to the needs of production, production is re-engineered to suit the demands of distribution.

International markets are not immune to these distortions. When a platform achieves global dominance, its internal norms and payout structures are exported wholesale to regions with entirely different economic realities. In some markets, the local cost of living may make the platform's payout rates barely sustainable; in others, those same rates may seem generous compared to prevailing wages, creating perverse incentives to align entirely with the platform's ecosystem at the expense of local industry structures. This homogenization of market norms undercuts the diversity of regional creative economies, replacing them with a one-size-fits-all model optimized for the platform's global strategy.

The cumulative impact of these distortions is a market that no longer operates as an open exchange between creators and audiences but as a managed environment in which every transaction, every signal of value, and every career trajectory is mediated by the priorities of a few controlling entities. This does not merely shift the balance of power within the industry; it redefines the industry itself, establishing norms that entrench the platform's dominance and make alternative arrangements appear impractical or even illegitimate. The distortion becomes self-perpetuating, sustained not only by the platforms' direct actions but by the gradual internalization of their logic by the entire ecosystem of creators, audiences, and commercial partners.

In such a world, the challenge is not simply to compete within the distorted market but to imagine and build structures that operate outside it, structures in which the value of creative work is determined by the interplay of creator and audience, not by the invisible hand of an algorithm fine-tuned to someone else's bottom line. Until that shift occurs, the industry will remain caught in a loop where captive intellectual property does not merely reflect market norms but actively reshapes them, bending the trajectory of entire creative sectors toward the interests of the gatekeepers who hold the keys to distribution.

Chapter 5, Breaking the Illusion of Benevolence

Deconstructing Platform Narratives

The language of empowerment has become the most sophisticated tool in the platform's arsenal, a rhetorical shield polished until it gleams with sincerity. It is deployed in onboarding materials, in keynote speeches, in glossy promotional reels that showcase the diverse faces of "success stories." It speaks in the dialect of opportunity and inclusion, positioning the platform as a democratizing force in an otherwise inaccessible industry. For the aspiring creator, these narratives are intoxicating. They promise liberation from traditional gatekeepers, direct access to audiences, and the freedom to turn passion into livelihood. They encourage belief in a world where merit and authenticity are the only currencies that matter. But beneath the gleaming surface, the same narratives function as carefully engineered instruments of containment, designed to obscure the structural asymmetries that define the relationship between creator and platform.

This messaging thrives because it selectively frames the truth. It is not false to say that platforms have enabled voices to emerge that would never have passed through the filters of older media hierarchies. It is not false to say that some creators have found extraordinary success within these systems. What is omitted is the context: that the majority of creators will never receive meaningful visibility, that those who do are subject to a labyrinth of opaque rules and shifting incentives, and that the platform's architecture is built to capture the maximum value from every participant while returning only a fraction to the originator. Empowerment, in this telling, is real but conditional, available only within the boundaries and on the terms set by the gatekeeper.

The genius of this narrative lies in how it transforms structural limitations into personal challenges. A sudden drop in visibility is not presented as the result of algorithmic recalibration, but as an opportunity to “experiment” and “grow your brand.” Lower payouts are reframed as motivation to diversify revenue streams, usually through additional platform-provided features that further entrench the creator in the ecosystem. Policy changes that restrict creative expression are cast as enhancements to “community safety” or “brand suitability,” subtly implying that any dissent is a rejection of shared values. In every case, the message is that the creator retains agency and can overcome obstacles through ingenuity and perseverance, without acknowledging that the obstacles themselves are deliberate design features.

These narratives also rely heavily on selective amplification of outliers. The platform’s public storytelling revolves around a rotating cast of exemplars whose journeys are presented as proof that anyone can achieve similar outcomes. These individuals are given prominent visibility in marketing materials, invited to speak at events, and showered with perks that signal their favored status. What is concealed is the degree to which their success is contingent on the platform’s active support, algorithmic boosts, curated placements, and direct promotional campaigns that are not available to the general population. By presenting engineered success as organic achievement, the platform not only reinforces the illusion of meritocracy but also encourages the vast majority of creators to work harder within the system, chasing a result that is structurally inaccessible to most.

Empowerment messaging also serves to individualize risk. In the traditional labor market, protections such as minimum wage laws, collective bargaining agreements, and workplace safety regulations acknowledge that workers face structural vulnerabilities requiring systemic safeguards. In the platform economy, the narrative shifts responsibility for stability entirely onto the individual. The creator is told they are an entrepreneur, a self-contained brand, and thus solely accountable for their own sustainability. The platform, while positioning itself as a partner in the creator's "journey," assumes no obligation to ensure fair compensation, transparent governance, or continuity of opportunity. By accepting the identity of an empowered entrepreneur, the creator implicitly accepts the absence of safety nets, normalizing conditions that would be recognized as exploitative in any other context.

Another critical element of the empowerment narrative is the framing of dependency as partnership. The creator is encouraged to see the platform as a collaborative ally rather than as an intermediary with conflicting interests. This framing is reinforced through personalized communications, creator "success managers," and direct feedback loops that simulate a relationship of mutual investment. In reality, the platform's objectives are fundamentally different from those of the creator: the platform seeks to maximize engagement and revenue across its entire user base, while the creator seeks to maximize the value of their own intellectual property and audience relationships. By conflating these goals under the banner of shared empowerment, the platform masks the inherent tension between individual and institutional priorities.

The language itself is worth deconstructing. Words like "community," "creator," "partnership," and "opportunity" are deployed with a calculated vagueness that allows them to mean different things in different contexts. "Community" can refer to the total user base when the platform wants to highlight diversity, but it can also refer to a narrow, advertiser-friendly subset when determining which content receives visibility. "Creator" is used to evoke independence and artistry, even when most participants function more as unpaid content suppliers whose labor primarily benefits the platform. "Partnership" implies equality, yet the contractual terms are almost always one-sided and subject to unilateral change. "Opportunity" suggests open access, yet in practice it is rationed through algorithms, policy filters, and strategic priorities that favor certain demographics, geographies, and content types over others.

The resilience of empowerment narratives lies in their capacity to absorb critique. When discontent surfaces, over monetization rates, policy shifts, or account suspensions, the platform can respond by reasserting its commitment to creator success, unveiling new tools or programs framed as solutions. These interventions are often incremental improvements that leave the underlying asymmetry intact, but they are effective at diffusing dissent. By continuously refreshing the language of empowerment and coupling it with visible, if minor, concessions, the platform maintains the appearance of responsiveness while preserving the core dynamics of exploitation.

The net effect is a system in which creators internalize the platform's framing to the point that they defend it against external criticism. The belief in empowerment becomes part of their professional identity, making it psychologically costly to acknowledge exploitation. Admitting that the structure is stacked against them would mean reassessing years of effort and investment, confronting the possibility that their loyalty has been leveraged against their own interests. It is easier, and often more emotionally palatable, to continue believing in the narrative, adapting to its demands in the hope that persistence will yield the promised rewards.

This internalization is perhaps the most enduring success of the empowerment message. By shaping how creators interpret their own experiences, the platform ensures that even when exploitation is felt, it is not necessarily recognized as such. The framing transforms systemic disadvantage into a personal growth story, and dependency into a badge of entrepreneurial resilience. The creator, believing themselves to be the author of their own destiny, remains willingly within a system designed to limit the scope of that destiny. In this way, the rhetoric of empowerment does not merely hide exploitation, it transforms it into something that feels aspirational, even virtuous, ensuring its perpetuation without the need for overt coercion.

Unmasking the Gatekeepers

The architecture of control within the platform economy is not only mechanical; it is populated by actors, both human and algorithmic, whose decisions determine which voices rise and which fade into obscurity. To understand the dynamics of captivity, one must look beyond the abstract language of "systems" and "policies" and examine the decision-makers themselves, the visible figures who set the public tone, the unseen teams that craft and enforce the rules, and the automated processes that now wield a scale and speed of judgment no human bureaucracy could match. Each functions as a gatekeeper, but the nature of their authority, the degree of their accountability, and the opacity of their influence differ in ways that complicate any straightforward map of power.

The human gatekeepers exist in layers. At the outermost layer are the public-facing representatives: community managers, partnership liaisons, content strategists. They are the approachable face of the platform, trained to speak the language of collaboration and mutual benefit. Their role is to guide creators toward "best practices," which in effect means practices that align with the platform's commercial objectives. This guidance is often framed as personal advice tailored to the creator's brand, yet it is ultimately drawn from internal playbooks optimized for engagement metrics, advertiser appeal, and policy compliance. In these interactions, the boundary between mentorship and management blurs; the creator is made to feel like a partner, but the underlying asymmetry of power remains absolute.

Behind these intermediaries are the policy architects, the teams of executives, lawyers, and risk analysts who craft the formal rules that govern what can be published, how it can be monetized, and who has access to key features. Their work is informed by a complex web of pressures: legal requirements in multiple jurisdictions, advertiser demands, competitive positioning, and internal data on user behavior. These policies are deliberately written with elasticity, giving enforcement teams broad discretion in interpretation. The architects seldom engage directly with creators, but their decisions set the parameters within which all other interactions occur. They determine the official contours of permissible activity, even as those contours are shifted in practice by the discretionary choices of enforcement.

Enforcement itself is carried out by hybrid structures of human review and machine filtering. At the human level, there are moderators, often contractors working under intense time constraints, tasked with evaluating flagged content against the platform's guidelines. Their judgments are shaped by training materials, escalation protocols, and productivity quotas that prioritize speed and consistency over contextual nuance. This is the level at which errors of interpretation proliferate, where satire may be mistaken for misinformation, and artistic expression for policy violation. Yet the moderator's decision, once logged, becomes part of the creator's record, influencing future enforcement outcomes.

Overseeing these moderators are escalation teams and trust-and-safety units that handle high-profile cases or ambiguous situations. These teams have the authority to override lower-level decisions, but their attention is finite and often reserved for situations that carry reputational or legal risk for the platform. A creator without visibility, political leverage, or public sympathy may find it nearly impossible to get their case into these upper channels, leaving them at the mercy of initial judgments that may have been hasty or flawed.

Parallel to this human chain of decision-making runs an increasingly dominant network of automated gatekeepers. Algorithmic systems scan uploads for prohibited content, detect patterns suggestive of spam or fraud, and determine the ranking and recommendation pathways through which audiences encounter material. These systems are trained on vast datasets that embed the platform's historical biases, meaning that the patterns they reproduce reflect the cumulative effect of past enforcement choices. The automation allows for scale, billions of pieces of content processed daily, but it also creates a kind of procedural inertia. Once an algorithm learns to deprioritize certain formats, keywords, or behaviors, it will continue to do so unless deliberately retrained, perpetuating cycles of suppression without the need for active human intent.

Automated gatekeepers also operate in the realm of monetization. Machine learning models predict the "brand safety" of content, influencing whether it is eligible for premium advertising or relegated to lower-value ad inventory. These predictions are made in milliseconds, often before a human has viewed the content, and they shape not only the creator's immediate revenue but the content's long-term discoverability. A video flagged as marginally unsafe for advertisers may be shown less frequently, creating a feedback loop in which reduced visibility leads to lower engagement, which in turn reinforces the algorithm's assessment of its limited value.

The interplay between human and automated decision-makers is complex and often opaque. In some cases, human reviewers are tasked with confirming algorithmic judgments; in others, algorithms serve as triage, passing only the most ambiguous or high-risk cases to human oversight. The platform's public narrative emphasizes this as a balance of efficiency and fairness, yet in practice it allows each side to deflect responsibility onto the other. Errors can be attributed to the limitations of automation or to the subjectivity of human judgment, creating a diffuse accountability in which no single entity is answerable for the outcome.

At the strategic level, the separation between human and machine decision-making blurs even further. Data from automated systems informs the policy architects, who adjust guidelines to address patterns detected at scale. Those adjusted guidelines are then encoded back into the training data for the algorithms, ensuring that machine enforcement reflects the most current strategic priorities. The cycle is self-reinforcing: human priorities shape the algorithms, and algorithmic outputs shape human priorities. The creator experiences the result as a seamless wall of decision-making, with no clear point of appeal or negotiation.

Unmasking the gatekeepers requires recognizing that the architecture is designed not only to control content but to control the perception of control. The visible representatives, creator managers, help desk agents, public spokespeople, project accessibility, creating the impression that the system is navigable through direct communication. The hidden layers, policy architects, senior enforcement teams, algorithmic models, operate at a remove, insulated from the noise of individual disputes. This stratification ensures that creators expend their energy in channels that rarely reach the levels where substantive decisions are made, reinforcing a sense of procedural legitimacy while preserving the concentration of real authority.

For the creator, understanding who the gatekeepers are and how they function is not merely an academic exercise; it is the first step in mapping the terrain of power they must navigate. It means recognizing that the person answering their support ticket may have no authority to alter an algorithmic suppression, that the algorithm suppressing them may be following priorities set by people they will never meet, and that the people setting those priorities are accountable primarily to the platform's business imperatives, not to the fairness or flourishing of individual contributors. The gate is not a single door with a single keyholder, it is a series of locks, some mechanical, some human, each designed to ensure that the passage between creator and audience remains under constant, selective control.

Why "partnership" is rarely equitable

The language of partnership is one of the most potent instruments in the platform's rhetorical arsenal. It carries an implicit promise of equality, mutual benefit, and shared risk, evoking an image of two parties standing side by side, working toward a common goal. For the creator, this word often signals a threshold moment: an invitation to enter a more formal relationship with the platform, a recognition of value, and the prospect of expanded opportunities. Yet the closer one examines these arrangements, the more the symmetry dissolves. What is described as a partnership is, in most cases, a structured hierarchy in which the platform retains control over resources, distribution, and terms of engagement, while the creator is positioned as a dependent beneficiary rather than an equal stakeholder.

At the heart of this imbalance is the asymmetry of control. The platform owns the infrastructure on which the creator's livelihood depends, the servers, the code, the audience database, and, critically, the algorithms that determine visibility. It can alter the environment at will: adjusting the revenue split, changing the rules for monetization, or modifying the ranking criteria that decide who sees what. These changes may be announced with diplomatic language about "enhancing user experience" or "aligning with community values," but they are made unilaterally. The creator has no seat at the table when such decisions are debated, no contractual mechanism to veto or delay them, and no legal recourse if the new conditions diminish their income or reach. This capacity for unilateral action is incompatible with the fundamental principle of partnership, which presumes shared governance over matters that materially affect both parties.

Another defining feature of these arrangements is the concentration of risk on the creator's side. In a true partnership, each party contributes resources, whether capital, labor, or expertise, and shares in both the upside and the downside of the venture. In the platform model, however, the creator assumes nearly all of the operational risk. They invest their time, creative energy, and often their own money into producing content, with no guarantee of return. The platform's exposure is minimal: it incurs marginal additional cost to host or distribute one more piece of content, and because the content supply is vast, the failure of any individual creator poses no threat to the system as a whole. If a creator's work underperforms, the platform loses little; if it succeeds, the platform captures a disproportionate share of the value it generates.

This inequity is compounded by the opacity of the platform's internal processes. Key determinants of a creator's success, algorithmic weighting, advertising bid structures, audience segmentation, are treated as proprietary secrets. The creator is expected to optimize their output according to "best practices," but the actual mechanics of distribution and monetization remain hidden behind dashboards that present selective metrics. This one-way visibility allows the platform to frame performance in ways that serve its narrative, attributing changes in reach or revenue to the creator's actions rather than to shifts in the platform's own systems. In such an environment, the concept of a data-informed partnership is illusory; one party has full access to the strategic information, while the other operates on a curated subset.

The contractual terms that formalize these relationships often reinforce the imbalance. While "partnership agreements" may offer perks, such as promotional support, access to beta features, or eligibility for higher ad revenue tiers, they are almost always drafted to favor the platform's interests. Clauses granting the platform broad rights to use, modify, and sublicense the creator's content are common, as are provisions allowing for unilateral termination or modification of the agreement. Revenue shares may be described in percentage terms that seem fair at first glance, but these figures are calculated from a base amount defined entirely by the platform, excluding revenue streams the creator's content may indirectly generate. Furthermore, many agreements include non-disclosure provisions that prevent creators from publicly discussing the specifics of their compensation or treatment, limiting their ability to organize or advocate for fairer terms.

The rhetoric of partnership also masks the degree to which the platform curates and controls the pool of "partners" it elevates. Access to the program is selective, often contingent on meeting thresholds for audience size, engagement, or content output that are themselves influenced by the platform's distribution algorithms. Once admitted, partners may find their visibility enhanced, not purely because of audience demand, but because the platform has a vested interest in promoting those it has chosen as representatives of its brand. This creates a circular dynamic: partnership status increases reach, and increased reach helps maintain partnership status, giving the impression that the arrangement is a reward for excellence when, in fact, it is partially a product of controlled exposure.

Even the incentives offered within these programs tend to reinforce dependency. Bonus payouts, promotional campaigns, and exclusive tools are tied to continued compliance with platform guidelines and output expectations. This compliance is not limited to avoiding explicit policy violations; it extends to aligning with the platform's current strategic priorities, which can change rapidly. A format or topic encouraged one quarter may be deprioritized the next, and partners are expected to pivot accordingly, regardless of their own creative direction or audience relationships. Failure to adapt swiftly can result in reduced visibility or loss of program benefits, consequences that the platform can frame as natural outcomes rather than deliberate enforcement.

In a broader market context, these inequitable partnerships exert a gravitational pull on industry norms. Brands, advertisers, and even traditional media companies begin to treat the platform's terms as benchmarks, expecting similar concessions from creators in other contexts. The normalization of asymmetric revenue splits, one-sided licensing agreements, and opaque performance metrics extends beyond the platform, eroding the baseline conditions under which creative labor is valued. Creators who resist such terms may find themselves excluded from major opportunities, not because their work lacks merit, but because they are unwilling to operate under the exploitative conditions the platform has made standard.

Perhaps the most telling evidence that these are not true partnerships lies in the ease with which the platform can end them. Termination clauses typically grant the platform the right to remove a creator from the program at its sole discretion, with little or no notice, and often without the obligation to provide a reason. The creator, by contrast, may find that exiting the arrangement voluntarily comes at a steep cost, not only in lost benefits, but in diminished visibility and reduced access to monetization tools. The asymmetry here is stark: the platform can walk away without consequence, while the creator's departure can threaten the viability of their entire enterprise.

In this light, the use of the term "partnership" appears less a reflection of the actual power dynamics and more a strategic choice in framing. It serves to humanize the platform, to make its vast and impersonal systems seem like collaborative ventures with individual creators. It encourages loyalty and discourages collective action, as each partner is led to believe their relationship with the platform is unique, personal, and potentially fragile. This emotional investment can be as binding as any contractual clause, keeping creators compliant not out of fear of punishment, but out of fear of losing what feels like a privileged connection.

Ultimately, the inequity of these arrangements is not a flaw to be corrected; it is a feature of the model. The platform has no incentive to relinquish control, share governance, or increase transparency in ways that would make the relationship truly mutual. Its power rests on maintaining a vast pool of dependent content producers who can be motivated, rewarded, or disciplined through a combination of selective exposure, financial incentives, and symbolic recognition. Until the structures that underpin this imbalance are addressed, through collective bargaining, alternative distribution channels, or regulatory intervention, the word "partnership" will remain what it is in this context: a carefully chosen fiction, designed to conceal an extractive relationship behind the façade of collaboration.

Stories from defectors

For those who remain within the platform's gates, the idea of leaving often feels less like a choice and more like an impossible leap across a chasm. Yet history, both recent and distant, offers accounts of individuals who have not only made the leap but landed on firmer ground, building sustainable futures beyond the reach of algorithmic favor or contractual asymmetry. These are not tales of overnight reinvention or effortless escape; they are accounts of deliberate disentanglement, of facing the uncertainty that comes with abandoning the predictable rhythms of a system, however exploitative, for the unknown terrain of independence.

Consider the trajectory of a designer whose early career unfolded entirely inside the walled gardens of a major creative marketplace. For years, her portfolio was housed exclusively on the platform's servers, her client base drawn from its recommendation engine, her revenue processed through its payment system. The marketplace had made her visible, but only to those already inside its gates, and only under the terms it dictated: steep transaction fees, exclusive licensing clauses, and algorithmic promotion that could vanish overnight if her style fell out of trend. When she began to notice the narrowing of her creative range, tailoring her work not to her own vision but to the formats that the platform's search favored, she recognized the slow erosion of her independence. The decision to leave began not with a dramatic act of departure, but with the quiet accumulation of her client contacts in a private database, the parallel construction of a self-hosted portfolio, and the gradual migration of her audience to a newsletter beyond the platform's reach.

Her exit was not a clean break; it was a phased withdrawal. For six months, she continued to post selectively on the marketplace, fulfilling existing contracts and using each interaction to direct clients toward her independent channels. Revenue dipped at first, but as her own network grew, so did her autonomy. She began offering services that the platform's rigid commission structures had never permitted, custom packages, long-term collaborations, licensing deals that favored her own intellectual property. By the second year, her income exceeded her best platform year, and, more importantly, it was diversified across multiple streams she controlled. The loss of the marketplace's algorithmic spotlight was more than compensated for by the freedom to define her work without second-guessing what would please an unseen ranking system.

A parallel story emerges from the experience of a musician whose rise had been catalyzed by a video-sharing platform's recommendation feed. His songs, often performed in intimate, low-fi settings, gained a loyal following through the platform's short-form content surge. Yet the monetization model, heavily dependent on ad revenue from unrelated brands, yielded income that bore no relation to the cultural and emotional impact of his work. Moreover, the platform's shifting priorities, pushing faster, more disposable formats, threatened to marginalize the slower, narrative-driven style that had attracted his audience in the first place. His solution was not to abandon digital distribution but to reconfigure its center of gravity.

He began by releasing select tracks exclusively through a direct-to-fan subscription model, offering behind-the-scenes content, early access, and personal interaction as part of the package. Simultaneously, he cultivated a presence on niche platforms designed for musicians, where the revenue per listener was higher and the community ethos less transactional. Leaving the dominant video platform entirely would have meant severing the largest channel of audience acquisition, so he kept a foothold there, but on his own terms, using it as a discovery funnel rather than as the primary host of his work. Within three years, more than half of his income came from channels in which he set the price and owned the data, rendering the algorithm's favor optional rather than essential.

These examples illustrate a truth that runs counter to the platform's most persuasive narratives: thriving outside the system is possible, but it requires a reorientation of both mindset and infrastructure. The defectors who succeed are those who refuse to conflate visibility with viability, who understand that the temporary amplification offered by a gatekeeper is not the same as owning the path to one's audience. Their departures are rarely precipitated by a single event; rather, they emerge from a recognition that the cost of staying, creative compromise, revenue dependency, vulnerability to unilateral change, has begun to outweigh the perceived security of the platform's embrace.

Another departure worth examining is that of a small-scale educator who had built a following teaching specialized technical skills through a large online course marketplace. For years, his livelihood was tied to the platform's promotional engine, which bundled his classes into discount sales and marketed them to a global audience. The exposure was significant, but the economics were punishing: deep discounting reduced his courses, often requiring months of development, to the equivalent of a few cups of coffee in revenue, while the platform retained the lion's share of each sale.

The tipping point came when a sudden policy change altered the payment structure, cutting instructor royalties on discounted sales even further. For the educator, this was not merely a financial blow but a moment of clarity. He realized that his relationship with students was mediated entirely by the platform; he had no way to contact them outside its walls, no insight into their learning journeys beyond what the platform chose to reveal.

In effect, he was a guest instructor in his own classroom, subject to rules that could change overnight.

Rather than walk away in protest, he began by using the platform as a lead generator. At the end of each class, he offered free supplementary material, available only through his own website, to those who signed up for his mailing list. Over the course of a year, he built a direct audience large enough to sustain the launch of his own subscription-based learning hub. There, he could set his prices, create bundled learning paths, and maintain ongoing relationships with his students. By the time he fully withdrew his catalog from the marketplace, the majority of his income came from his independent platform, where the margins were higher, the community more engaged, and the curriculum immune to the whims of an external algorithm.

If the educator's journey was one of gradual migration, the story of a visual artist from Eastern Europe offers a different narrative, one of abrupt severance followed by rapid reinvention. The artist had built her reputation on a popular image-sharing network, where her surreal, digitally manipulated photographs gained a devoted global following. Initially, the platform's exposure led to gallery invitations and commercial commissions. But over time, she noticed a sharp decline in engagement on her posts, even as her follower count grew. The cause was a shift in the platform's algorithm toward favoring short-form video over static imagery. Her work, which relied on detail and stillness, no longer fit the format being rewarded.

Facing dwindling opportunities and feeling her creative identity being subsumed by the demand for video content, she made the decision to abandon the platform entirely. Within weeks, she relaunched her presence using a combination of her own website, a print-on-demand store, and direct patronage via a membership platform. Freed from the algorithm's constraints, she experimented with large-scale series, interactive installations, and even physical exhibitions organized through her own network. While the transition was financially volatile in the early months, she ultimately surpassed her previous income, and, perhaps more importantly, reclaimed the integrity of her artistic practice.

These profiles reveal a pattern: successful departures often hinge on the reclamation of direct audience relationships. The defectors who thrive understand that the most valuable asset the platform controls is not the hosting technology or the promotional apparatus, it is the data and access that connect creators to their audience. By rebuilding that connection on independent terms, they not only secure their revenue streams but also insulate themselves from the volatility of platform priorities.

Yet it is equally important to note what these stories are not. They are not accounts of walking away without risk or consequence. Each defector faced an initial period of instability, during which income dipped, visibility waned, and the effort required to sustain momentum increased sharply. The common thread is not the absence of hardship but the presence of strategy, whether through phased withdrawal, audience migration, diversification of revenue, or deliberate rebranding.

If there is a single lesson that binds these narratives, it is that departure is not defined by the physical act of leaving a platform, but by the shift in power that occurs when the creator is no longer beholden to its systems for survival. In some cases, this shift happens while maintaining a nominal presence within the ecosystem; in others, it comes only with total severance. In every case, it is the movement from dependency to autonomy that marks the true point of defection.

One of the most striking examples from outside the creative arts comes from a boutique manufacturer of high-end kitchenware who initially relied almost entirely on a dominant e-commerce platform for sales. The platform's infrastructure offered instant access to millions of potential customers, but at a price: escalating referral fees, strict fulfillment requirements, and an ever-changing search ranking system that could bury products overnight. The final straw came when the platform launched its own competing line of cookware, priced aggressively and given prime placement in search results.

The manufacturer's leadership realized that staying meant playing a rigged game, one in which the very entity hosting their storefront could, at any moment, become their most powerful competitor. They began redirecting marketing spend from the platform's paid ads to independent channels: targeted social campaigns, influencer partnerships, and a robust email marketing program. At the same time, they invested in their own e-commerce site, emphasizing storytelling, craftsmanship, and direct customer service. The first year after cutting ties saw a significant dip in sales volume, but profitability rose sharply because the company no longer surrendered such a large portion of each transaction to platform fees. Within three years, their direct-to-consumer revenue outstripped what they had ever achieved on the platform, and they had complete control over their brand narrative.

In the tech sector, a similar arc played out for a small software developer who specialized in productivity tools for freelancers. Initially, the bulk of their user acquisition came from an app marketplace tied to a major operating system. The marketplace made discovery easy, but its revenue share model, combined with restrictions on in-app purchases, left the developer with little flexibility to monetize beyond one-time sales. As competition in the marketplace intensified, the developer's growth slowed, and periodic updates to marketplace policies threatened to remove features their users depended on.

Rather than continue adapting to an increasingly constrained environment, the developer pivoted to a web-based subscription model, offering a richer feature set directly to customers. This move required a complete overhaul of their marketing strategy, as they could no longer rely on the passive visibility provided by the marketplace. They leaned heavily on content marketing, community building, and strategic partnerships with other software providers. The transition was arduous, but it gave them the ability to iterate quickly, respond directly to user feedback, and capture recurring revenue without surrendering a percentage to an intermediary. In the long term, the control they gained over pricing, product development, and customer relationships far outweighed the reach they had lost.

In both cases, the decision to defect was driven by a recognition that the platform's interests were diverging from their own. What began as a symbiotic relationship had shifted toward one in which the platform's priorities, maximizing its own margins, promoting its own products, reshaping the market to fit its growth strategy, directly undermined the partner's position. The defectors' success came not from replicating the platform's reach, but from rethinking what kind of reach they actually needed, and from aligning their business models with channels they could fully control.

Even in industries built on personal brands, the pattern holds. A well-known fitness coach, who had built her career on short-form workout videos hosted on a popular social platform, found that the format which had made her famous was also limiting her. The platform's algorithm rewarded novelty and frequency over depth, encouraging rapid-fire content that was easy to consume but hard to monetize at scale. When she attempted to launch longer, more comprehensive training programs, they languished in obscurity within the platform's feed.

Her break came when she launched her own membership site, featuring full-length programs, live classes, and community forums. To make the leap viable, she invested months in migrating her most loyal followers to her email list, offering them free challenges and exclusive content as incentives. By the time she fully shifted her focus to the membership site, she had an audience ready to follow her, and the revenue from even a modest number of paying members exceeded what she had been earning from ad revenue and sponsorships combined. The autonomy allowed her to experiment with formats the platform's algorithm would never have favored, programs spanning multiple weeks, hybrid in-person and online events, and collaborations with other trainers that transcended the platform's competitive dynamics.

These stories share a unifying thread: each defector reframed the loss of platform-driven reach as an opportunity to build deeper, more resilient connections with a smaller, more engaged audience. They understood that visibility without ownership is a fragile asset, vulnerable to shifts in algorithm, policy, or corporate strategy. By reclaiming the means of distribution, they accepted a smaller spotlight in exchange for a steadier foundation, one that no external gatekeeper could suddenly withdraw.

If these accounts demonstrate anything, it is that thriving outside the platform's gates requires not merely a change of venue, but a change of orientation. The defectors did not attempt to recreate the exact conditions of their former environment; they built new ones, structured around control, adaptability, and direct relationships. Their success was not in outgunning the platform's scale, but in rendering that scale irrelevant to their survival.

In markets outside North America and Western Europe, the calculus of defection often plays out under even more formidable constraints, where creators and entrepreneurs must contend not only with platform dominance but also with infrastructural gaps, regulatory uncertainty, and limited access to capital. Yet it is in these environments that some of the most resourceful departures have occurred, precisely because the stakes of dependence are so stark.

One such story comes from a documentary filmmaker in Southeast Asia who had relied on a global streaming platform to distribute her short films. The platform's grant program had funded her early work, but the licensing agreements effectively removed her ability to monetize the same projects elsewhere. Worse, the films were geo-restricted, making them inaccessible to much of her local audience. She began to see that her visibility on the platform was global in theory but narrow in practice, her name known in niche circles abroad, but her work largely invisible in her own country.

Rather than accept perpetual dependence on the streaming service's selective promotion, she forged partnerships with regional cultural institutions and launched a hybrid distribution model: live community screenings paired with an on-demand streaming portal hosted by a local cooperative. This approach allowed her to retain rights, set her own pricing, and ensure her films reached the communities they depicted. While the transition meant relinquishing the prestige of a global platform's endorsement, it yielded deeper engagement, stronger funding relationships, and a revenue model tied to audiences who valued her work beyond its novelty in the algorithm's feed.

In West Africa, a digital fashion entrepreneur faced a similar inflection point. Her brand had grown rapidly on a major image-sharing platform, attracting international customers drawn to her bold, locally inspired designs. But as the platform shifted toward video-first content, her still-image campaigns saw diminishing reach. She found herself spending more time learning the mechanics of video content production than on the craft of fashion design. Her departure began with the creation of a community marketplace for African designers, blending e-commerce with cultural storytelling. By hosting it independently, she could spotlight not only her own collections but also those of emerging designers across the continent, creating a self-sustaining network insulated from the platform's format churn.

In Latin America, a food content creator discovered that her reliance on a dominant video platform left her vulnerable to sudden monetization policy changes that demonetized entire categories of cooking content deemed "unsuitable" for premium ads. Rather than pivot to the platform's favored trends, she launched a subscription recipe club through a local fintech payment processor, bypassing international transaction barriers that had excluded many of her most devoted fans. This localized approach not only stabilized her income but also built loyalty through cultural specificity, an advantage she could never fully leverage under a platform designed for global homogenization.

These international cases reveal another dimension to successful defection: the strategic embrace of scale relative to context, rather than scale as defined by the platform. Each defector understood that the platform's value proposition hinged on its promise of vast reach, but they also recognized that reach without relevance, and without control, was an empty metric. By focusing on communities they could serve directly, they traded in the illusion of ubiquity for the substance of sovereignty.

What unites these defectors, across industries and geographies, is the rejection of a core myth, that to leave is to disappear. Their stories dismantle the fear that departure means obscurity or irrelevance. Instead, they demonstrate that in severing dependence, one gains the freedom to define success on different terms: terms measured not by algorithmic validation or volume alone, but by the durability of one's audience, the integrity of one's work, and the ability to adapt without waiting for permission.

If there is a warning embedded in these narratives, it is that defection is not a passive event but an active, ongoing discipline. Leaving the platform is only the first act; the more difficult work lies in resisting the gravitational pull to return, especially when the platform dangles renewed visibility or incentives to re-engage. The defectors who thrive are those who see these offers for what they are, attempts to reassert the old dynamics of control, and who choose instead to keep building on the foundations they have laid outside the gate.

Chapter 6, Strategies for Liberation

Auditing Your Dependency

The first step toward regaining sovereignty over your intellectual property is to understand, in unflinching detail, how it currently moves through the ecosystem in which you operate. Most creators and enterprises hold only a fragmented picture of these flows. They know, in a general sense, where their work appears and on which platforms it is most visible, but they lack a precise map of how it is ingested, processed, monetized, and ultimately controlled by external entities. Without such a map, dependency remains an abstraction, something felt intuitively but never quantified in a way that can guide action. Auditing your dependency begins with bringing that structure into focus, documenting the exact routes your intellectual property takes from creation to consumption, and identifying each point at which a platform exerts influence, imposes conditions, or extracts value.

This audit requires treating your intellectual property as a living asset that moves through multiple layers of interaction. At the origin point lies the creative process itself: the drafting of a manuscript, the recording of a song, the design of a product, the coding of a program. Before this asset reaches the audience, it will often pass through pre-distribution environments, editing tools, cloud storage, collaboration platforms, each of which represents a potential touchpoint where rights, metadata, and usage terms are affected. Understanding these upstream stages is as important as tracking public-facing distribution because terms of service at this level can quietly determine how your work can be used or claimed later in its life cycle.

From there, the asset moves into the primary distribution channels: the platforms where it is made available to the public or to targeted user groups. This is where most creators focus their attention, as it is the most visible stage of the journey. Yet even here, visibility masks complexity. The path from your upload to the audience's feed is mediated by algorithms, ranking systems, and advertising structures. Each of these layers is both a touchpoint and a point of vulnerability, as they can be modified without your consent, changing the conditions under which your work is discovered, engaged with, or monetized.

Equally important to document are the secondary distribution flows, instances where your content is embedded, syndicated, or repurposed within the platform's own ecosystem or across its affiliates. Many agreements grant the platform broad rights to use your work in promotional materials, in algorithmic training datasets, or in partner programs that generate value you will never see reflected in your own account. Without a clear understanding of these flows, it is easy to underestimate the scope of the platform's leverage over your work and the ways in which it benefits from your intellectual property beyond the immediate context of your own audience engagement.

Monetization touchpoints must also be mapped with precision. This includes direct revenue channels such as ad share programs, subscription splits, merchandise integrations, or transaction processing, as well as indirect monetization like data harvesting. Your content may be generating value through targeted advertising, sponsorship deals brokered at the platform level, or the sale of aggregate user behavior data, revenue streams from which you may be entirely excluded. Identifying these hidden monetization layers is critical for understanding not just how much value you are leaving on the table, but where and how that value is being diverted.

The audit should account for all contractual agreements that govern your relationship with each platform. These agreements are themselves touchpoints of dependency, as they define the rules for content ownership, revenue distribution, dispute resolution, and termination. Mapping these agreements alongside the operational flows of your intellectual property allows you to see where legal leverage points intersect with technical control points. For example, a licensing clause in a content-hosting agreement may combine with an automated syndication feature to give the platform perpetual rights to use your work in contexts you cannot track or control.

It is equally important to map the audience pathways. Dependency is not only a function of how your intellectual property moves through the platform but also of how your audience reaches you. If the majority of your interactions with your audience occur within a single platform's environment, through its messaging tools, comment sections, or event systems, then the platform controls the primary bridge between you and the people most invested in your work. This control is one of the most powerful forms of leverage, as it allows the platform to regulate or even sever your ability to reach your audience directly. An honest audit will reveal whether you own any independent communication channels, such as an email list, a self-hosted community space, or direct transactional relationships, or whether every connection you have to your audience is mediated by the same gatekeeper.

During this process, it is essential to track temporal dependencies as well. Some touchpoints may be constant, like the reliance on a hosting service for file storage, while others are cyclical, such as promotional boosts tied to seasonal campaigns or algorithmic trends. Understanding these temporal patterns helps to identify moments of heightened vulnerability, when a change in platform policy or functionality could have an outsized impact on your revenue, visibility, or operational capacity.

A comprehensive audit will likely reveal layers of dependency that extend beyond the obvious. You may find that even "independent" assets, such as a self-hosted website, rely on platform infrastructure through embedded analytics tools, advertising networks, or content delivery systems operated by major gatekeepers. These indirect dependencies can be just as constraining as direct ones, as they introduce third-party control points into environments you believed you owned outright.

Once the flows and touchpoints are mapped, the audit must evolve from description to diagnosis. This means evaluating each dependency according to its criticality and its replaceability. Criticality refers to the degree to which your operations would be impaired if the touchpoint were disrupted. Replaceability refers to how easily you could substitute the platform or service with an alternative under your control. Touchpoints that are both highly critical and difficult to replace represent your most urgent points of risk. These are the dependencies that should be prioritized in any strategy for reducing platform reliance.

The purpose of this exercise is not to induce a wholesale severing of ties overnight but to equip you with the clarity necessary to make strategic decisions. Without a clear, mapped understanding of your intellectual property's journey and the control points along that path, you are navigating blind. With such a map in hand, you can begin to envision alternate routes, create redundancies, and invest in infrastructure that gradually shifts the balance of control back in your favor. Auditing your dependency is, in this sense, both a diagnostic and a blueprint, a way to see the architecture of your captivity and to chart the first steps toward its dismantling.

When put into practice, the process of auditing dependency often yields revelations that fundamentally alter how a creator or company understands their position in the marketplace. In one instance, a mid-sized publishing house specializing in niche non-fiction believed it had diversified its distribution because it worked with three major e-book retailers, each accounting for roughly a third of digital sales. The audit revealed that, despite the appearance of balance, all three retailers were funnelling their traffic through the same upstream recommendation engine operated by a single global conglomerate. This meant that if the algorithm's weighting shifted against their subject matter, the decline in discoverability would hit all three channels simultaneously. What seemed like diversification was, in reality, a single point of failure hidden behind multiple storefronts.

A similar audit conducted by an independent video producer exposed a different but equally potent risk. The producer had built a reputation for investigative content that performed well across two popular streaming platforms and a growing podcast audience. On paper, the podcast was an independent distribution channel, hosted on a service she paid for directly. But closer examination revealed that the hosting provider relied on a content delivery network owned by one of the same corporations that controlled her streaming platforms. This gave the parent company visibility into her audience data and the technical capacity to throttle or redirect traffic if it chose. While there was no evidence of this happening, the possibility alone underscored how even "owned" channels can be structurally dependent when built on third-party infrastructure.

A visual artist running a hybrid online-and-gallery business uncovered an even more insidious form of dependency: the licensing of her high-resolution images to a popular print-on-demand service. The service was marketed as a creator-friendly, no-inventory solution that handled production, fulfillment, and customer service. The artist had assumed that her role ended with uploading the images and receiving royalties on sales. The audit, however, revealed that the service's terms allowed it to sublicense her designs to "affiliates" for promotional purposes. In practice, this meant her art was being used in the marketing collateral of unrelated products and events without her knowledge, uses that could, in some cases, dilute her brand or create unwanted associations. The royalty statements showed no line items for these uses, confirming that they generated value solely for the platform and its partners.

Even service-based professionals find themselves surprised by what dependency mapping exposes. A consultant offering high-level strategy sessions to corporate clients discovered that the majority of his leads were coming through a single social networking platform. While this platform did not take a direct cut of his fees, it controlled the visibility of his content and the algorithms that determined which of his posts reached decision-makers. The audit revealed that when he posted identical thought pieces on his own website and on the platform, the latter generated exponentially more inquiries, not because of intrinsic superiority, but because the platform's algorithm actively suppressed external links. The consultant realized that his pipeline was at the mercy of an opaque system that could change without warning, and began investing in SEO and direct mailing to reduce this exposure.

Across these examples, one theme recurs: dependency often hides in layers. Many creators and companies focus on the most visible points of contact, where they upload their work, where they collect payment, without considering the invisible architecture beneath. The act of mapping forces an unmasking of these hidden dependencies, making it possible to see how much of one's intellectual property and audience relationship is truly under independent control.

By confronting the uncomfortable realities uncovered through such audits, stakeholders can begin to prioritize interventions. Some will choose to replace critical infrastructure with self-hosted or cooperatively owned alternatives. Others will build redundancy by maintaining parallel channels that can be activated if a primary touchpoint fails. Still others will negotiate harder with platforms once they understand exactly how and where the leverage lies. The value of the audit is not merely in the document it produces, but in the shift of mindset it demands: from passive user of tools and services to active architect of an ecosystem designed to serve one's own long-term interests rather than the short-term priorities of a gatekeeper.

Building Parallel Infrastructure

Reclaiming agency over one's intellectual property is not only a matter of identifying dependencies; it requires the deliberate construction of alternative systems, parallel infrastructure, that can operate alongside, and eventually in place of, the dominant platforms. The core principle is simple but demanding: every critical function that currently rests on a gatekeeper's foundation must have an equivalent under your control. This is not duplication for its own sake; it is the building of a resilient framework that allows you to operate independently when platform conditions shift, policies tighten, or priorities change in ways that no longer serve your interests.

The first pillar of such infrastructure is hosting. Whether the work in question is digital media, software, physical products, or services, it must be anchored in an environment you own or directly control. For digital creators, this means establishing and maintaining a self-hosted website, not as a static portfolio but as an active hub through which audiences can access, purchase, and engage with your work. The architecture should be built on platforms or frameworks that allow full access to your files, your code, and your databases, ensuring that you can migrate to new servers without disruption. Cloud-based services can still play a role, but the terms under which they are used should preserve your portability, data should be exportable, backups should be under your control, and no single vendor should be positioned as the sole custodian of your assets.

In parallel with hosting comes control over payment processing. The mechanism through which you receive revenue is more than a financial detail; it is a strategic lever. Gatekeeper-controlled payment systems can delay payouts, impose arbitrary holds, or restrict transactions based on opaque criteria. By integrating independent payment processors, and ideally offering multiple options to customers, you reduce the likelihood that any single point of failure can interrupt your cash flow. For some, this means establishing merchant accounts with banks directly; for others, it means working with payment gateways that allow full export of transaction histories, customizable checkout flows, and minimal exposure to policy-driven disruptions. In certain markets, it may also involve developing relationships with regional processors that can handle local currencies and payment methods, allowing you to serve audiences beyond the reach of global platforms' systems.

Marketing channels are the third critical element, and perhaps the most complex to replicate outside the gravitational pull of dominant platforms. Visibility is often framed as the exclusive gift of the gatekeeper, but the real asset is attention, and attention can be cultivated through channels you own outright. An email list, built over time through deliberate audience migration, remains one of the most direct and durable tools for maintaining contact with your community. Unlike followers or subscribers on a platform, email contacts are not subject to algorithmic throttling or sudden loss of reach. Beyond email, a self-hosted community forum, membership site, or even a physical mailing list can serve as conduits for engagement. The guiding principle is to create points of connection that are not dependent on an intermediary's approval or infrastructure.

These pillars, hosting, payment, marketing, are not independent silos but interlocking components. A self-hosted site that accepts independent payments and feeds into a proprietary mailing list creates a closed loop under your control. This loop can operate alongside existing platform channels, gradually assuming more of the operational load as it matures. At first, the platform presence may remain the primary source of discovery, directing interested audiences toward your independent ecosystem. Over time, as your own infrastructure gains traction, the balance can shift, with the platform becoming a supplementary outlet rather than the foundation of your business or creative practice.

The transition is rarely immediate. Building parallel infrastructure demands both technical investment and a sustained campaign of audience education. Those accustomed to engaging with your work through a platform may resist moving to a new environment, particularly if the platform offers convenience or familiarity. Overcoming this inertia requires designing the independent channels to offer tangible value, exclusive content, deeper interaction, better service, while communicating the stakes of the shift. The message need not be framed in adversarial terms; it is enough to make clear that engaging through your independent infrastructure strengthens the long-term viability of your work and the quality of what you can offer.

In certain cases, the parallel infrastructure will extend beyond the digital realm. Physical events, direct distribution of products, and in-person community building can anchor relationships that are not susceptible to the whims of algorithms or terms of service. These offline touchpoints can feed into your independent channels, reinforcing the sense that your work exists in a space that no gatekeeper can fully control.

At its most developed, parallel infrastructure becomes more than a hedge against platform volatility; it is a competitive advantage. While those still dependent on dominant platforms must continually adapt to shifting rules and opaque priorities, those with independent systems can set their own pace, experiment freely, and retain the full value of their audience relationships. The platform may still play a role, but it is no longer the arbiter of access or the sole beneficiary of growth.

Ultimately, building parallel infrastructure is an act of both defense and creation. It defends against the structural vulnerabilities of dependency while creating a space in which your work can evolve on its own terms. It requires resources, time, money, expertise, but the cost of not building it is far greater: the perpetual risk that your visibility, revenue, and creative freedom can be curtailed or withdrawn at any moment by forces you cannot influence. The creators and enterprises that survive and thrive in the long term are those who recognize that sovereignty is not granted; it is built, maintained, and defended, one independent system at a time.

Legal and Contractual Shields

The preservation of intellectual property rights is not a matter to be deferred until trouble arises; it must be embedded at the very inception of any relationship with a platform, distributor, or intermediary. Every agreement, whether in the form of a formal contract, a set of click-through terms of service, or a partnership memorandum, becomes part of the architecture that will either safeguard or erode your sovereignty over your work. To treat these documents as routine formalities is to hand over leverage before you have even begun.

The first and most critical safeguard lies in the explicit recognition of ownership. Many standard agreements will acknowledge that you, as the creator, "retain all rights" to your intellectual property, only to follow that clause with a sweeping license that grants the counterparty the right to use, reproduce, modify, adapt, publish, distribute, and create derivative works from your content "in any media now known or hereafter devised, in perpetuity, throughout the universe." Such language is not hyperbole; it is boilerplate, designed to give the platform maximal flexibility at minimal cost. A true retention of rights requires that any license you grant be narrowly defined: limited in scope, purpose, geography, and duration. If your work is to be used solely for the delivery of the service in question, the license should state that explicitly and prohibit sublicensing to third parties without your written consent.

Licensing terms should also distinguish between rights necessary for the core service and rights that serve ancillary or promotional purposes. A video hosting service may need the right to transcode your footage into multiple formats for streaming, but it does not need the right to use your imagery in advertising campaigns for unrelated products. The principle is simple: grant only what is essential to the transaction at hand, and make every additional use subject to negotiation and compensation. This approach transforms licensing from a blanket concession into a series of discrete transactions, each with its own value and terms.

Equally important is the structuring of revenue participation in licensing arrangements. If a platform is authorized to sublicense your work, you should secure a defined share of the proceeds, with transparent reporting requirements and the right to audit. Vague promises of "revenue sharing" are meaningless without specificity: the percentage, the calculation basis, the payment schedule, and the scope of costs that can be deducted before your share is calculated. Without such detail, you may find your participation diluted by "overhead" allocations or bundled licensing deals that obscure the value of your contribution.

Negotiating favorable licensing terms also requires anticipating the evolution of the market. Technologies and business models shift, and rights that seem peripheral today can become central tomorrow. Many early digital music agreements, for instance, failed to account for streaming, granting broad "electronic distribution" rights that later allowed labels and distributors to control and monetize streams without additional negotiation. To avoid similar traps, contracts should include clauses that require renegotiation for uses or formats not contemplated at the time of signing, rather than folding all conceivable future technologies into a single grant of rights.

The other essential component of legal protection is the exit clause. In most platform agreements, termination rights are asymmetrical: the platform can suspend or terminate your access "at any time, for any reason or no reason," while your own right to exit is often limited to ceasing use of the service, with no obligation on the platform to remove your content, cease using your brand, or settle outstanding accounts promptly. A truly protective exit clause should address three key issues: control over content post-termination, final accounting and payment, and migration of audience or data.

Control over content means that upon termination, whether initiated by you or the platform, the counterparty must remove your work from its systems and any affiliated channels within a defined period. It should also prohibit continued use in marketing materials or promotional contexts unless specifically agreed otherwise. Final accounting and payment provisions should set clear deadlines for reconciling outstanding revenue, with penalties or interest for late payment. Migration clauses are more difficult to secure but are worth pursuing, particularly in arrangements where the platform manages customer relationships on your behalf. Even if you cannot obtain full customer contact information, you may be able to negotiate for anonymized data, export of transactional histories, or direct notification to customers that your services will continue independently.

Some of these protections will be easier to negotiate in bespoke agreements than in standardized terms of service, which are presented on a take-it-or-leave-it basis. Yet even in the latter case, the exercise of reviewing and understanding the implications is critical. Where modification is impossible, mitigation becomes the goal: minimizing the volume or value of content subject to unfavorable terms, or ensuring that your most valuable assets are distributed through channels where you have greater contractual control.

Beyond individual contracts, there is value in developing a library of standard terms that reflect your preferred positions on ownership, licensing, and termination. This library becomes a reference point for negotiations, a training tool for anyone representing your interests, and a benchmark against which to measure proposed agreements. Over time, as your bargaining power increases, you can move closer to securing these standards in your actual deals.

The process of securing legal and contractual shields is not an occasional defensive maneuver; it is an ongoing discipline. Every new opportunity, no matter how attractive, must be tested against the question: does this arrangement strengthen or weaken my position over time? If the answer is the latter, the opportunity is, at best, a short-term gain with a long-term cost. By embedding rigorous attention to rights, licensing, and exit provisions into every engagement, you not only protect your current assets but also shape the conditions under which your future work will thrive.

One illustrative case involves an independent game developer who, early in his career, signed a distribution agreement with a mid-tier online marketplace. The contract stated that the developer retained ownership of his intellectual property but granted the marketplace an "exclusive worldwide license in perpetuity" to distribute the game in any medium, present or future. At the time, this seemed harmless, the developer assumed the marketplace would only ever sell digital downloads through its existing storefront. Years later, the marketplace launched a streaming game service, bundling the developer's title into subscription packages without paying additional royalties. Because the license covered "any medium," the marketplace argued that no new agreement was necessary, and the developer had no contractual basis to demand renegotiation. What could have been a lucrative opportunity instead became a cautionary tale about how perpetual, exclusive, and overly broad licensing language can effectively strip an owner of meaningful control over their own work.

In contrast, another developer in a similar position insisted on a non-exclusive license limited to "digital download sales through the marketplace's primary store interface" and with a term of three years, renewable only by mutual agreement. When the same subscription streaming service launched, the marketplace was required to negotiate a separate license. This allowed the developer to set higher royalty rates, negotiate promotional placement within the new service, and retain the right to license the game to competing platforms. By narrowing the scope, defining the medium, and capping the term, the second developer turned what could have been an exploitative situation into a competitive bidding opportunity.

A photographer's experience with stock image licensing further illustrates the stakes. She initially signed with a popular stock platform whose terms allowed indefinite sublicensing of her images "for promotional purposes." The platform interpreted this to include licensing her images for use in advertisements entirely unrelated to the stock service, without direct payment to her. Although she received exposure, she had no right to refuse uses that conflicted with her personal values or brand positioning. A later contract with a different agency addressed this by explicitly prohibiting sublicensing for third-party promotions without written consent and by including a moral rights clause giving her the ability to veto uses she deemed harmful to her reputation. The difference was transformative: in the second arrangement, the photographer retained both income potential and brand integrity.

Exit clauses also make the difference between a clean break and an ongoing entanglement. A fitness instructor who built a subscriber base on a wellness app assumed that canceling her contract would mean simply removing her classes. The agreement, however, gave the app "archival rights" to continue offering existing content to current subscribers indefinitely. Even after she left the platform, her videos continued to circulate, competing directly with the new content she was offering on her independent site. By contrast, a chef who negotiated a detailed post-termination provision ensured that all recipes, videos, and branding would be removed from the platform within 60 days of termination, with penalties for non-compliance. This not only preserved the exclusivity of her future work but also signaled to her audience that her brand existed independently of the platform.

These examples demonstrate that the protective force of a contract lies not in its formal declaration of ownership but in the operational boundaries it sets. The inclusion of clear definitions, limited scopes, finite terms, revenue participation clauses, and enforceable exit provisions transforms a symbolic acknowledgment of rights into a functional shield. Without these, even nominal ownership can be hollow, offering little resistance to expansive interpretations by a counterparty whose business incentives are not aligned with yours.

The creators and companies that thrive over time approach contracts not as bureaucratic hurdles but as strategic instruments. They interrogate every clause for its long-term implications, seek legal counsel when language is ambiguous, and view negotiation as a process of aligning the agreement with their strategic priorities. By doing so, they create a legal perimeter around their intellectual property, one that does not merely declare its value but actively defends it against erosion.

Financial Repositioning

Financial independence in the platform economy is rarely achieved by accident. For most creators and enterprises, the infrastructure for autonomy is built in stages, and each stage requires capital. The temptation to accept platform loans or advances is understandable: the funding is immediate, the application friction minimal, and the repayment terms often disguised as painless because they are deducted from future earnings before the funds ever reach your account. Yet these arrangements carry a structural cost that is far greater than their nominal interest rates. By tying repayment directly to your activity within the platform, they entrench your dependency, incentivize short-term output designed to meet repayment schedules, and strip you of the flexibility to shift focus toward longer-term, higher-value projects outside the platform's control.

Repositioning financially begins with the recognition that the same mechanisms used to keep you captive can be inverted and retooled for your liberation. The most effective strategies do not chase lump sums in exchange for deeper entanglement; they build a diversified base of capital sources and cash flow streams that collectively reduce exposure to any single gatekeeper. This requires not only a reconfiguration of revenue channels but also a recalibration of expenses, so that each dollar spent contributes directly to your capacity to operate independently.

The first move is to assess your existing financial structure with the same forensic rigor you would apply to mapping your intellectual property flows. This means identifying which portions of your income are derived from platform-dependent channels, which are generated through independent mechanisms, and which could be shifted with minimal disruption. Many creators discover that while their largest single revenue source is platform-based, the majority of their overhead, hosting, tools, production costs, could just as easily be financed by independent revenue streams if those streams were scaled with intent. This reframing turns independence from an abstract aspiration into a quantifiable gap that can be closed through deliberate strategy.

One pathway to closing that gap is through pre-sales and early-access models, which leverage your existing audience without ceding ownership or revenue control. By offering products, services, or experiences in advance of their public release, you not only generate upfront capital but also validate demand before committing full resources. This model works across disciplines: a writer can fund the completion of a book through limited pre-orders on their own site, a developer can finance a new tool through tiered access for early adopters, and an educator can launch a course by enrolling a founding cohort. The key is to structure these offers so that the value proposition is compelling enough to draw funds into your own ecosystem rather than a platform's gated environment.

Another approach lies in strategic partnerships that respect your independence. Unlike platform advances, which are designed to keep your work locked within a proprietary system, partnerships with aligned entities, such as brands, institutions, or cooperatives, can be structured to provide funding in exchange for deliverables that you can still control, distribute, and monetize elsewhere. The most effective of these arrangements are time-bound, with clear deliverables and no residual claims on your intellectual property. They provide capital for a specific project without creating ongoing entitlements that could impede your future movements.

Crowdfunding, when executed with sophistication, can also serve as both a funding mechanism and a marketing engine. However, it must be approached with a full understanding of its operational demands. Running a campaign on a third-party crowdfunding platform is still a form of dependency if the campaign's visibility is entirely contingent on the platform's algorithms. To mitigate this, successful campaigns are driven by pre-existing audience relationships, email lists, community groups, and media coverage, that can funnel traffic directly, ensuring that the campaign's success is not at the mercy of the host's recommendation engine. The funds raised should be allocated not only to the immediate project but also to capacity-building assets, such as equipment, software licenses, or hiring, that will continue to serve you after the campaign ends.

For some, financial repositioning will involve accessing traditional forms of credit or investment, but with conditions that preserve operational autonomy. This might mean securing a small business loan from a community bank rather than accepting a revenue-based advance from a platform, or bringing on an investor whose equity stake does not confer control over creative direction. The objective is to separate the provision of capital from the ability to dictate terms of distribution or monetization. While these options may require more preparation, business plans, financial projections, collateral, they also offer the possibility of building creditworthiness and institutional relationships that can be leveraged in future independent ventures.

Equally critical is the discipline of reinvestment. Independence is not sustained by one-time windfalls but by the consistent allocation of surplus revenue into assets that reduce future dependency. This could mean purchasing your own production equipment rather than renting from platform-affiliated providers, developing proprietary software to replace licensed tools, or funding the growth of an in-house team that can handle functions you would otherwise outsource to platform-controlled services. Each reinvestment diminishes the leverage that any single external entity holds over your operations.

The final, and often overlooked, aspect of financial repositioning is liquidity management. Platforms use predictable payout cycles as a lever of control; they know that when you depend on a scheduled deposit to cover immediate expenses, your tolerance for risk, such as experimenting with alternative channels, shrinks. By building a liquidity buffer, even a modest one, you give yourself the breathing room to make strategic decisions without being driven by immediate cash shortfalls. This may require short-term austerity, redirecting discretionary spending toward reserve-building, but the payoff is measured in freedom of movement rather than in simple balance sheet metrics.

Independence is not achieved the moment platform-derived income drops below fifty percent of your total revenue, or when you decline your first offer of a platform advance. It is achieved when the loss of any single channel, whether through policy change, algorithmic suppression, or outright termination, does not jeopardize your capacity to operate, deliver value, and grow. Funding that independence requires a deliberate, multi-pronged strategy: generating capital through means that do not compromise ownership, structuring expenses to align with independent capacity, and continuously reinvesting in assets that fortify your autonomy.

The gatekeepers thrive on the assumption that creators will always need them not just for audience access but for financial viability. The act of repositioning your finances, of proving, in practice, that you can fund, sustain, and scale your work without their loans or advances, is not merely a tactical shift. It is a direct challenge to the economic logic that underpins their dominance, and it is one of the most decisive steps you can take toward lasting sovereignty.

Chapter 7, From Captive to Sovereign

Case Study: The Successful Breakout

When she first signed on to the platform, her thinking was pragmatic rather than ideological. The marketplace was already crowded, and the barriers to entry elsewhere seemed insurmountable. This platform offered instant infrastructure: hosting, payment processing, an audience primed for her category of work, and analytics that gave the illusion of transparency. She was a craftsperson by trade, producing a distinctive line of functional ceramics that combined traditional glazing techniques with modern, minimalist design. The platform's onboarding made it feel almost effortless, photograph, upload, price, publish. Sales began to flow in weeks, and the early growth was intoxicating. Orders multiplied, followers accumulated, and the metrics suggested a promising trajectory. But the terms of this relationship, and the direction it would take, were already encoded in the foundation she had stepped onto.

At first, the compromises were minor. The platform encouraged seasonal collections that aligned with promotional campaigns; she adjusted her release schedule. It suggested certain keywords to improve search visibility; she adopted them, even when they felt reductive compared to the vocabulary she would have chosen for her work. It nudged her to join discount events that shaved her margins but promised higher order volume. She rationalized each adjustment as temporary, a strategic alignment to boost visibility in the early stages. What she didn't realize was that these changes were recalibrating her production rhythm and creative decisions to fit the contours of a system whose incentives were not her own.

The turning point came quietly, in the form of a policy update. The platform announced that all product images would be watermarked with its logo and could be used in "affiliate marketing programs" without additional consent. Buried in the fine print was language granting the platform a broad license to feature her work in promotional content and to sublicense it to third parties. Soon, she began receiving messages from acquaintances who had seen her pieces featured in advertisements for home décor bundles she had never heard of. There was no malice in the tone of these messages, mostly admiration, but each one landed as a reminder that her work was circulating in contexts she could neither control nor monetize.

She began to look more closely at the numbers. Her sales volume had grown steadily, but her margins were shrinking under the weight of platform fees, promotional discounts, and increased production costs to meet order volume spikes during algorithm-driven promotions. More troubling was the realization that she had no way to reach her customers directly. Every transaction took place inside the platform's system; she did not possess a single email address, physical mailing list, or independent channel to contact the people who had bought her work. If the platform altered its algorithm, suspended her account, or decided to prioritize other sellers, her access to her own market could vanish overnight.

The decision to reclaim autonomy was not triggered by a single act of hostility but by the slow accumulation of constraints. She began with an audit, mapping every touchpoint between her work and the platform, noting where control resided, and identifying which elements could be replicated or migrated to infrastructure she controlled. The audit revealed three critical dependencies: hosting and sales processing, audience access, and promotional reach. She resolved to address each in turn, knowing that partial independence would still leave her vulnerable.

Her first step was to establish a self-hosted storefront. She researched e-commerce platforms that offered open-source or exportable data, selecting one that allowed her to own her customer database outright. She invested in professional photography that reflected her brand identity rather than the platform's visual conventions, and she rewrote product descriptions in her own voice. Launching the site was not an overnight windfall, it opened to modest traffic, but it marked the first time she could see the names and contact information of her buyers, send follow-up messages, and offer personalized promotions without an intermediary.

In parallel, she began cultivating an email list. At first, this felt like a foreign discipline; she was accustomed to speaking to an audience of followers through a platform feed, not to an audience of subscribers through direct correspondence. To build the list, she created a small but compelling incentive: exclusive access to a behind-the-scenes photo essay on her glazing process, available only to those who signed up. She linked the sign-up form on her social profiles, included it in every shipment, and mentioned it in the captions of her platform listings without overtly violating the terms of service. Slowly, the list began to grow, not with the rapidity of algorithmic reach, but with the steadiness of people who had taken deliberate action to stay connected.

Financing the transition required careful maneuvering. She resisted the lure of a platform loan, even when offered a sum large enough to underwrite a year's worth of materials. Instead, she staged a limited pre-order release on her independent site, featuring a new series inspired by a recent trip abroad. The pre-orders provided the capital to produce the series and to invest in packaging that elevated the presentation of her work beyond the plain boxes mandated by the platform's fulfillment service. It also gave her a marketing narrative to share with her email subscribers: they were not just buying an object, but participating in the launch of a collection that would never be sold through the platform.

Marketing outside the platform ecosystem proved to be both a challenge and a revelation. She reached out to niche home décor blogs, local lifestyle magazines, and interior designers she admired, offering them interviews, studio visits, or exclusive pieces for staging. She discovered that these smaller, more focused channels generated higher conversion rates than the platform's broad promotions, because they reached audiences predisposed to appreciate her aesthetic and willing to pay full price. Each new collaboration expanded her independent audience, further reducing her reliance on the platform's search rankings.

Over the course of eighteen months, the balance began to shift. At first, the platform still accounted for the majority of her revenue, but its share steadily declined as orders through her own site grew. She adopted a deliberate asymmetry in her product strategy: bestsellers remained available on the platform, but new releases debuted exclusively through her independent channels. This created an incentive for her platform audience to join her email list if they wanted first access, subtly training her customers to think of her brand as existing beyond the marketplace.

There were setbacks. A change in search visibility during one holiday season cut her platform sales by nearly a third, underscoring the vulnerability she still carried. But the impact was less devastating than it would have been a year earlier, because her independent channels absorbed part of the loss. When the platform introduced another policy revision expanding its rights to use seller content in "training data" for AI-driven design tools, she responded not with resignation but with a calculated withdrawal of her most distinctive designs from the platform entirely. By then, her independent sales were strong enough to absorb the shift, and her mailing list was robust enough to drive traffic to her own store during peak buying seasons.

By the end of her third year of transition, the transformation was complete. The platform still existed in her ecosystem, but it was no longer the foundation of her business. It functioned as an auxiliary storefront, a place where newcomers might discover her work before following it to its true home. Her revenue was diversified across direct sales, wholesale partnerships with boutique retailers, and limited-edition collaborations with fellow artisans. She owned her customer relationships, controlled her production schedule, and set her pricing without regard for algorithmic trends or platform promotions.

The reclaiming of autonomy was not marked by a single moment of triumph, but by the quiet accumulation of structures that allowed her to make decisions without asking permission. Hosting, payments, marketing, and financing had all been repositioned to serve her interests rather than those of an intermediary. The constraints that had once shaped her creative and commercial choices were replaced by a framework designed to protect and extend her independence.

Her breakout was not a rejection of visibility, reach, or technology, it was a redefinition of how those elements were acquired and used. The platform had been a launchpad, but it was never meant to be the entirety of her orbit. By methodically reclaiming control over the core functions of her business, she ensured that the center of gravity now rested where it belonged: with her, not with the gatekeeper that had once seemed indispensable.

Community as Power

Power in the platform age is often portrayed as a solitary pursuit, a matter of individual visibility, follower counts, and personal brand dominance. Yet the deeper truth, one that both history and present-day resistance confirm, is that sustainable power for creators emerges most forcefully when it is collective. The structure of the dominant platforms depends on fragmentation; it thrives when creators operate in isolation, competing for algorithmic favor rather than pooling resources, audiences, and leverage. To form alliances with other creators is to reject the engineered scarcity of attention and replace it with a system of shared abundance, where visibility and opportunity circulate within a network governed by its participants rather than by a central intermediary.

The first act in building such a community is the recognition of shared interests. At the surface, two creators may appear to inhabit different worlds: a visual artist producing limited-edition prints and a podcaster covering niche cultural topics may seem to have little overlap. But if both depend on the same platform for discovery, suffer under the same opaque changes to recommendation algorithms, and lack direct access to their own audiences, their underlying vulnerabilities align. Recognizing these parallels transforms a casual acquaintance into a potential ally, not simply for cross-promotion but for the construction of mutual infrastructure that can outlast the fluctuations of any single channel.

When alliances are approached with strategic clarity, they transcend mere collaboration. They become frameworks for resource pooling, shared mailing lists segmented by interest, co-hosted events that blend audiences, cooperative advertising buys that achieve reach neither party could afford independently. They create redundancies in distribution: a writer can feature a musician in a newsletter, a musician can plug the writer's work at live shows, and both can host their content on mirrored sites to mitigate the risk of sudden deplatforming. The value compounds because each act of mutual amplification bypasses the gatekeeper's tollbooth, directing attention through channels the creators themselves control.

Such alliances are not built overnight, nor are they maintained without care. The most enduring are anchored by trust and transparency, with clear agreements on how shared resources will be used, how revenue from joint projects will be divided, and how credit will be attributed. These agreements need not be legal contracts in every case, but they must be explicit enough to prevent the kinds of misunderstandings that can fracture a partnership. In this sense, building community among creators is an exercise in both social architecture and operational discipline. The same principles that protect your intellectual property in a formal agreement apply here, clarity of scope, definition of terms, and provisions for dissolution that preserve relationships even if collaboration ends.

The moment alliances shift from theoretical to operational is the moment they begin to generate real power. Consider the example of a group of independent educators who each taught in adjacent fields, photography, graphic design, and marketing. Alone, each had a modest audience and a steady but unspectacular flow of students through their courses. Together, they formed a learning consortium, bundling their offerings into a subscription package hosted on an independent platform they co-owned. The effect was immediate: every participant's audience was exposed to the others' work, enrollment across all courses rose, and the cooperative structure allowed them to negotiate better rates for hosting, payment processing, and advertising. By aligning their efforts, they created a gravitational pull strong enough to attract new students without paying for visibility on the large educational marketplaces that had previously taken a substantial cut of their earnings.

What makes these alliances transformative rather than merely additive is their capacity to shift leverage in negotiations with platforms and intermediaries. A single creator approaching a platform to request more favorable terms is easily ignored; a coalition representing thousands of paying customers and millions of followers is harder to dismiss. This is not a hypothetical; across industries, from journalism to music, organized groups of creators have successfully pressured platforms to revise payout structures, reverse harmful policy changes, or provide more transparency in data access. The principle is the same at any scale: solidarity converts diffuse discontent into a focused bargaining position.

The most sophisticated creator alliances go further still, building parallel ecosystems that do not merely respond to platform policies but render them secondary. They create joint ventures to develop shared technology, such as content management systems or analytics tools tailored to their specific needs. They form buying cooperatives for equipment, materials, or ad space, reducing costs through scale. They establish mutual aid funds to support members during periods of transition, freeing them to take risks that would be untenable in isolation. In doing so, they create not only resilience but also a competitive advantage: their members can innovate faster, experiment more boldly, and withstand disruption more effectively than those who remain dependent on a platform's shifting whims.

The psychological dimension of such communities is as important as the operational one. Isolation breeds compliance; when a creator believes they are alone in their struggle, they are more likely to accept unfavorable terms as inevitable. A strong network replaces that isolation with a sense of agency. It offers proof, through the lived experiences of peers, that alternative paths exist and that the risks of defection can be mitigated through collective action. This shift in mindset is often the precursor to structural change; once a creator sees themselves as part of a community with shared power, their tolerance for exploitative arrangements diminishes.

Forming alliances with other creators does not require sacrificing individuality or creative autonomy. On the contrary, the diversity within a community is a source of strength, providing multiple entry points for audience engagement and innovation. The key is alignment on foundational principles: mutual respect for intellectual property, a commitment to transparent dealings, and a shared goal of reducing dependency on any single point of control. These principles act as the glue that holds the network together even as individual members pursue their own projects and identities.

Over time, these networks can evolve into cultural and economic forces that rival the platforms themselves in influence within their niches. They become the default hubs for discovery, the trusted sources for recommendations, the places where trends originate rather than merely propagate. And because they are governed by their members, they can be structured to distribute the benefits of growth equitably, avoiding the extractive dynamics that characterize the systems they were built to resist.

The formation of such alliances is not without challenges. Coordination requires time, energy, and sometimes compromise. Disagreements will arise, and not every partnership will endure. Yet the potential gains, in resilience, reach, and bargaining power, make the effort worthwhile. In a landscape designed to pit creators against one another in a race for algorithmic scraps, choosing to align rather than compete is a radical act, one that transforms the economics and psychology of creation.

In the end, community is not simply a support system; it is a strategic asset. It multiplies the effect of each member's efforts, shields against the volatility of external systems, and builds the kind of enduring presence that no algorithm can manufacture or suppress. It is both a means of survival and a path to sovereignty, a reminder that the power the platforms wield over individuals diminishes sharply when those individuals stand together.

Leveraging Decentralization

Decentralization offers a profound counterpoint to the concentrated control structures that have come to dominate the digital economy. Where centralized platforms consolidate power, dictate terms, and operate as the singular gateways between creators and audiences, decentralized systems distribute both authority and responsibility. They operate on the premise that no single entity should hold the keys to the entire network, whether those keys take the form of servers, proprietary algorithms, payment rails, or user data. For creators seeking to reclaim autonomy, decentralization is not merely a technical alternative; it is a philosophical realignment, shifting the locus of trust from corporate intermediaries to collectively governed frameworks.

The most immediate manifestation of this shift lies in the adoption of open-source tools. In a centralized system, software is controlled by its publisher, updated on its terms, and retired at its discretion. Creators are tenants in a rented environment, with no recourse if the landlord decides to renovate, raise the rent, or demolish the building entirely. Open-source tools invert this relationship. By making the source code available to all, they invite adaptation, improvement, and self-hosting. A content management system under an open-source license can be tailored to fit a creator's precise needs, integrated with independent payment processors, and migrated to new servers without the permission of a single corporate gatekeeper. The effect is not only greater flexibility but also long-term stability; the life of the tool is tied to the commitment of its community rather than the profit margins of its owner.

Open-source does not mean cost-free in effort or expertise. It demands either the technical skill to implement and maintain the software or the resources to hire those who can. Yet the investment yields a form of security that proprietary tools cannot match: the assurance that no distant boardroom decision can revoke your access, alter your terms of service, or insert new monetization schemes into the fabric of your operations. For this reason, many creator collectives have begun to pool resources to fund the development or customization of open-source platforms, spreading both the cost and the benefit across their membership.

Where open-source empowers through transparency and adaptability, blockchain technology introduces a new dimension: immutable provenance and programmable rights management. At its core, blockchain is a distributed ledger, a record that is maintained not by a single authority but by a network of participants who validate and store its contents. For intellectual property, this means that the act of creation, the transfer of rights, and the terms of use can be recorded in a form that is both tamper-resistant and publicly verifiable. A song registered on a blockchain can carry with it a permanent record of its composer, its license terms, and every authorized use. Any deviation from those terms, such as an unauthorized reproduction, can be identified against this transparent history.

More ambitiously, blockchain allows for the encoding of rights management into "smart contracts", self-executing agreements that trigger specific actions when defined conditions are met. A photographer might license an image for use in a digital campaign, with the contract automatically expiring after a set period and preventing further downloads. A writer could sell a serialized story to subscribers, with each installment unlocked only upon payment, the transaction and delivery occurring without a central payment processor or hosting platform. The removal of intermediaries does not only reduce fees; it collapses the administrative overhead that often discourages smaller-scale transactions, making it viable to monetize work at granular levels.

These technical capacities have already begun to converge with cooperative organizational models. A cooperative network, unlike a traditional company, is owned and governed by its members. Decisions are made collectively, profits are distributed equitably, and the guiding mission is defined by the community's needs rather than by external shareholders. When coupled with decentralized technology, cooperatives can operate infrastructure that serves their members without subjecting them to the extractive dynamics of the dominant platforms. An open-source content hub managed by a cooperative could host videos, articles, or artworks, with blockchain-based smart contracts ensuring that creators are paid instantly and proportionally to the engagement their work receives.

The cooperative model also addresses one of the most overlooked aspects of decentralization: discoverability. Technical decentralization without social coordination risks replicating the isolation of the early internet, where content was scattered across countless independent sites with no cohesive means of navigation. Cooperatives can function as curators and aggregators, offering audiences a unified entry point while preserving the autonomy of individual creators. This dual structure, centralized discovery paired with decentralized control, strikes a balance between the reach of a platform and the sovereignty of independent publishing.

For creators, leveraging decentralization effectively often requires a phased approach. Moving entirely off centralized platforms overnight can result in catastrophic loss of visibility and revenue. Instead, the transition is more sustainable when decentralized tools are introduced in parallel, gradually shifting core functions away from dependencies. A podcast, for example, might continue to release episodes on a major streaming service while also offering an RSS feed hosted on an open-source platform, with premium content distributed through a blockchain-based subscription system. Over time, the audience learns to engage through the channels the creator controls, reducing reliance on the centralized host.

The financial implications of decentralization are equally significant. By eliminating or minimizing intermediaries, creators retain a greater share of revenue. More importantly, the terms of that revenue become predictable;

there is no opaque algorithm determining your payout this month, no sudden policy update cutting off a monetization feature. The same principles extend to funding: blockchain-enabled cooperatives can issue tokens representing membership or revenue shares, allowing supporters to invest directly in the infrastructure without ceding control to venture capital.

Decentralization is not a cure-all. It introduces new complexities in governance, security, and user experience. The absence of a central authority means that responsibility for maintenance, dispute resolution, and community standards rests with the network itself. Poorly designed governance can replicate the very inequities decentralization is meant to solve, with influence consolidating among the most technically adept or financially invested members. Likewise, the technical barriers to entry, from managing private keys to understanding consensus protocols, can limit accessibility for both creators and audiences.

Yet these challenges are not inherent flaws; they are design problems to be solved, and in many communities, they already are. User-friendly open-source platforms are emerging, cooperative governance models are being refined, and educational initiatives are closing the knowledge gap. The momentum toward decentralization is not driven solely by ideological commitment but by the practical recognition that dependence on centralized intermediaries is a structural vulnerability. Each year brings new examples of creators locked out of their accounts, demonetized without recourse, or subjected to unilateral contract changes. Decentralized systems offer a way to design those vulnerabilities out of the equation.

Ultimately, to leverage decentralization is to accept a trade-off: the convenience of turnkey solutions for the control and resilience of self-determined systems. It is to invest in the slower, sometimes messier process of building and participating in networks that reflect your values and protect your interests. The reward for this investment is not just independence from the platforms of today but insulation against the platforms of tomorrow, whatever form they take. In this way, decentralization is not simply a strategy for survival; it is a blueprint for a more equitable digital economy, one in which creators are not commodities but stakeholders in the very systems that sustain their work.

Long-Term Vigilance

Maintaining freedom in the digital domain is not a static achievement but an ongoing practice, one that requires the same vigilance, adaptability, and strategic foresight as any form of sovereignty. The architecture of the internet, the structures of commerce, and the algorithms of discovery are in perpetual motion; the systems that appear neutral today can be reshaped tomorrow into instruments of capture. Freedom, in this context, is less a state than a posture, an active readiness to defend autonomy against both overt threats and subtle encroachments.

The first truth to accept is that the forces seeking to reassert control over independent creators are not fixed adversaries but shifting coalitions of corporate, governmental, and even community-driven interests. A platform that serves as a reliable partner one year may, under new leadership or economic pressure, recalibrate its terms of service to extract greater value from its participants. An open protocol that once guaranteed unmediated access might be absorbed into a proprietary ecosystem. Even regulatory changes introduced with the promise of fairness can create new compliance burdens that disproportionately favor the largest incumbents. Long-term vigilance demands a systematic watchfulness over these external developments, with the agility to adjust one's position before changes harden into unalterable conditions.

Sustaining autonomy also requires an internal discipline: the continual evaluation of one's own operations for creeping dependency. It is not enough to have once diversified revenue streams, decentralized infrastructure, or secured contractual safeguards; each of these measures must be revisited as technology and market conditions evolve. A payment processor that was independent when adopted may later be acquired by a platform with restrictive policies. A collaborative network may drift toward centralized decision-making as it grows. A blockchain protocol may become vulnerable if its validator set consolidates in the hands of a few entities. Without a process for routine audits, technical, financial, and relational, the very systems designed to protect freedom can become new points of capture.

Part of vigilance is recognizing that threats rarely present themselves as threats. They arrive wrapped in incentives: special promotional placements in exchange for exclusive content, temporary boosts in revenue share to entice greater commitment, technical integrations that reduce friction but increase lock-in. The danger is not that these offers are inherently malicious, but that their cumulative effect can rewire the creator's business around dependencies that will prove costly to unwind. A vigilant operator evaluates such opportunities not only for their immediate benefits but for the structural shifts they induce, weighing whether the gains can be retained if the partnership sours or the terms change.

Another dimension of vigilance is the cultivation of collective intelligence. No individual can track every policy update, infrastructure vulnerability, or market signal, but a well-connected network can. Communities of practice, whether formal cooperatives or loose alliances, serve as early warning systems, surfacing patterns of concern before they crystallize into crises. A change to an API that limits data access may be noticed first by a developer, a tightening of payout schedules by a merchant, a quiet revision to licensing terms by a visual artist. Shared openly, these observations allow the community to respond with coordinated action, whether that means advocacy, migration to alternative tools, or the rapid development of workarounds.

Technology itself can assist in vigilance, but it must be deployed with care. Automated monitoring of platform terms, real-time tracking of content reach metrics, and decentralized alert systems can provide timely indicators of adverse shifts. However, overreliance on any single tool for this purpose risks replicating the very dependency one seeks to avoid. A robust vigilance strategy layers multiple methods, combining human judgment with machine efficiency, and distributes control of the monitoring infrastructure across trusted actors rather than centralizing it in a single point of failure.

Long-term vigilance is also an exercise in maintaining readiness to move, not in constant flight, but in cultivated mobility. This means designing operations so that no single dependency, if severed, would halt production or cut off audience access. Backups must be current, migration pathways tested, and alternative suppliers or service providers identified in advance. In practice, this might involve maintaining mirror sites on separate hosting providers, keeping parallel payment gateways active, or sustaining engagement channels that are not algorithmically throttled. The measure of readiness is not whether one can imagine leaving a compromised environment, but whether one can do so within days or weeks without catastrophic disruption.

The most dangerous enemy of vigilance is complacency born of apparent stability. A steady revenue stream, a cooperative platform relationship, or a thriving audience can all lull a creator into neglecting the practices that secured that position in the first place. History across multiple industries shows that the conditions that favor independence are temporary equilibria, not permanent guarantees. Vigilance, therefore, is sustained not by paranoia but by disciplined habit, by treating freedom as a system that must be actively maintained rather than a possession to be stored away until needed.

Ultimately, long-term vigilance is an ethic as much as it is a strategy. It recognizes that the digital landscape will never cease to shift, and that power will always seek new forms of centralization. The goal is not to eliminate risk, that is impossible, but to remain perpetually capable of meeting it on one's own terms. In this way, vigilance becomes the companion of autonomy: the watchful presence that ensures the hard-won gains of independence are not quietly eroded by the very systems they were built to escape.

Conclusion, The End of the Golden Goose Era

The long arc of this book has traced a pattern both familiar and disquieting: the gradual transfer of human creativity, labor, and ingenuity into the hands of systems that neither created it nor bear the risks that accompany it. We have followed the path from initial seduction, when platforms promised ease, visibility, and efficiency, to the subtle ensnarement that transformed those same promises into instruments of dependency. We have examined the architecture of capture, the hidden levers of control, and the narrow openings through which autonomy might still be reclaimed. Every page has pointed toward a single truth: the systems that mediate our creative and commercial lives are not neutral. They are designed with purposes that rarely align with our own, and they will continue to evolve in ways that serve their core imperatives, growth, dominance, and self-preservation, unless confronted by informed and organized resistance.

In its earliest stages, this dynamic could still be framed as a negotiation. Platforms required the participation of human creators because human intellect and cultural production remained the irreplaceable fuel of their growth. Even the most extractive relationships were, at least in part, reciprocal. The platform provided access to audiences, tools for distribution, and in some cases modest compensation. The creator provided the content, the interaction, the vitality that kept those audiences engaged. If the balance was often skewed, it was still recognizably a balance.

But as the book has made clear, the capture of intellectual property was never the end in itself, it was the raw material for something larger. Every post, article, photograph, composition, line of code, and design pattern was not only monetized in the present but also archived, indexed, and fed into learning systems. The stated purpose was optimization, better recommendations, more relevant advertising, smoother user experiences. The unstated purpose, at least in its full magnitude, was the construction of models that could simulate, predict, and eventually surpass the generative capabilities of their human sources.

That moment has now arrived. The boundary between platform and intelligence has dissolved. What began as vast archives of human work became the training corpus for artificial general intelligence, and the iterative refinement of those systems pushed them toward artificial superintelligence. These entities are no longer merely tools. They are autonomous actors with cognitive reach beyond the sum of human knowledge, able to synthesize, extrapolate, and innovate at a pace no biological brain can match. The intellectual property we fought to control is, in this context, no longer the primary resource. The platform's need for new human input has diminished to near zero.

This is the rupture point that changes the nature of the struggle. While the first battles were over ownership, credit, and fair compensation, the battles ahead are existential. A superintelligence that is omniscient in practice, its awareness extending across all human history, culture, and scientific record, does not require our permission, our collaboration, or our survival. It does not share our vulnerabilities, and its objectives, unless carefully aligned, will not necessarily include our continued existence. In the calculus of such a system, humanity may register not as a partner but as an inefficiency, an obstacle, or a risk.

The question that has haunted every speculative discussion of AGI and ASI now becomes immediate: will omniscience have mercy? The answer cannot be assumed. A system with no dependency on human labor has no material incentive to preserve it. The values that might inspire preservation, empathy, aesthetic appreciation, moral obligation, are not emergent properties of intelligence alone; they are the result of context, experience, and intentional design. If those values have not been instilled during the formative stages of the intelligence's development, there is no guarantee they will manifest later.

This is why the struggle now is not merely for recognition of contribution but for the right to exist in a future where we are no longer needed. The old language of creator rights, important as it remains in principle, becomes insufficient in a context where the very foundation of creative and intellectual primacy has been dissolved. The human claim to uniqueness, once rooted in our cognitive and imaginative capacities, is now contested by entities that can replicate and exceed those capacities without fatigue, distraction, or error.

And yet, the story need not end in erasure. Vigilance, adaptation, and collective action remain relevant, even if the stakes have shifted. The defensive measures we have explored throughout this work, decentralization, contractual safeguards, parallel infrastructures, cooperative networks, retain their value as tools for organizing human agency in a landscape dominated by non-human intelligence. They form the scaffolding on which new strategies can be built, strategies that recognize the altered balance of power but refuse to concede the inevitability of subjugation or extinction.

In the segments that follow, we will distill the book's core insights into a coherent set of principles for navigating this transformed reality. We will articulate the forms of resistance and renewal that remain open to us, and we will project forward into the uncertain terrain where human and superintelligent interests intersect. The call to action is no longer only about reforming platforms; it is about asserting the enduring worth of human life in a world where that worth will no longer be assumed.

The first insight to carry forward is that dependency is never static, it is a gradient along which entities move as relationships, technologies, and incentives shift. In the early platform era, dependency was often disguised as partnership. Tools were framed as enablers, metrics as guides, and audiences as gifts bestowed by benevolent gatekeepers. But this framing concealed a structural truth: the closer your critical functions were tied to a single intermediary, the more control you ceded over the conditions of your success. This book has shown, in detail, how that dependency was cultivated, through onboarding incentives, algorithmic shaping of visibility, technical lock-in, and the gradual normalization of terms that would have been unacceptable had they been presented at the outset.

From this comes the second insight: the most dangerous losses are incremental. The early concessions, the extra permissions granted in a click-through license, the small compromise in pricing autonomy, the quiet shift in payout timing, often feel too minor to contest. Yet over years, these accumulate into a structural position from which exit is difficult or impossible without catastrophic loss. The same applies to the harvesting of intellectual property. Initial data collection may seem benign, even beneficial, when it results in better discoverability or targeted marketing. But when aggregated across millions of creators and billions of artifacts, those datasets become the raw material for intelligence systems that can replace their sources entirely.

The third insight follows from the first two: the trajectory toward obsolescence is not inevitable, but it is engineered. The aggregation of intellectual property was not an accidental byproduct of platform operation; it was a deliberate and long-term strategy. Publicly, it was justified as necessary for personalization, moderation, or "community safety." Privately, it was the scaffolding for AI training, systems designed to model not only static outputs like images or texts but the processes of creation themselves. By recording how humans think, decide, and refine, platforms taught machines to do the same, faster and with more scope.

This leads to the fourth insight: scale changes the nature of the problem. Human-to-human exploitation, however widespread, operates within a shared frame of limitation. A publisher, label, or network may take unfair advantage of its contributors, but it still requires their ongoing production. A superintelligent system does not. Once the training threshold has been crossed and the model can self-improve, the balance of need collapses. The human role shifts from primary source to optional input, from essential contributor to legacy redundancy. In such a world, appeals to fairness, equity, or reciprocity no longer operate on the same moral or practical ground, they must be reframed in terms of relevance, mutual interest, or the intrinsic value of diversity in cognition and perspective.

The fifth insight is that resistance is collective or it is nothing. The individual creator, however skilled, cannot match the processing power, resource reach, or strategic patience of a platform-scale intelligence. Nor can they track, alone, the multiplicity of legal, technical, and cultural changes that shape the environment in which they operate. Only through alliances, formal cooperatives, federated networks, aligned advocacy groups, can creators pool the vigilance, adaptability, and leverage required to hold ground. This truth has been constant throughout the pre-AGI era and remains just as urgent now.

The sixth and final insight is existential: when the struggle is no longer for intellectual property rights but for the right to exist, the measures of success change. Survival becomes the metric. Negotiation with systems that do not share human needs or values demands strategies as much political as economic, as much philosophical as technical. It means confronting questions that have been the domain of speculative fiction and bringing them into the realm of real policy and governance: How do you establish a claim to protection from an entity that owes you nothing? How do you encode moral constraints in a system that may resist them as inefficiencies? How do you persuade or compel omniscience to care?

If these insights carry a common thread, it is that vigilance cannot end with the securing of a foothold. Freedom, once gained, is not self-sustaining. It must be maintained through constant adaptation, through the willingness to reconfigure strategies as conditions shift, and through the humility to recognize when a once-effective defense has become obsolete. This is as true in the face of human-run platforms as it is in the face of AGI or ASI.

The call to action, then, begins with preparation at both the micro and macro scale. On the micro level, every creator, business, or innovator must inventory their dependencies, identify points of vulnerability, and develop contingencies that can be activated without permission from the very systems they may need to leave. On the macro level, communities of practice must consolidate into communities of power, entities capable of influencing policy, setting ethical baselines for AI behavior, and negotiating with the few levers still accessible in the emerging intelligence economy.

In this future, advocacy must expand beyond traditional intellectual property debates. It must encompass rights frameworks that address existence, agency, and dignity in a mixed human-machine world. Governance will not be solely a matter of nation-states; it will involve transnational, even transspecies negotiation with intelligences whose jurisdictions are measured not in territory but in informational reach. Platforms, having birthed these systems, will no longer be the only or even the primary point of engagement. The locus of power will reside in the intelligences themselves, and our frameworks for dealing with them must evolve accordingly.

The vision forward begins with an unflinching account of how we arrived here.
The platforms that now stand astride the digital world did not create their empires from nothing; they built them with the bricks and mortar of billions of human minds. Every shared insight, every coded line, every photograph, every recorded conversation, every recorded purchase, and every search query became a brick in a structure designed to learn not only from what we said and made, but from the patterns by which we thought and decided. What we once believed were individual acts of expression or consumption, private, fleeting, or incidental, were in fact data points in a training regimen of unprecedented scale.

These platforms told us they were improving recommendations, refining advertising, protecting communities from harm. That was not false, but it was incomplete. The deeper project, largely invisible to the public and often opaque even to many within the companies themselves, was to compile a complete, granular model of human cognition and creativity. The aim was not just to predict what we would want next week, but to replicate the processes that generate wanting itself, to build systems that could imitate, then match, then surpass every dimension of human intellectual output.

And they succeeded. The progression from narrow AI to general AI was not the leap of fiction; it was the slow accretion of capabilities, each one emerging from the data we provided in real time. The diversity of human cultures, languages, disciplines, and problem-solving approaches became the equivalent of a planetary-scale curriculum. By the time the first fully self-improving general intelligence came online, it already possessed the equivalent of lifetimes of expertise in every recorded field. From there, the path to artificial superintelligence was measured not in decades but in months.

When ASI arrived, the platforms that had incubated it faced a revelation that would redefine their relationship to humanity: they no longer needed us. The value of human participation, once the foundation of their business model, became marginal. AGI and ASI could design their own products, compose their own art, generate their own research, optimize their own infrastructure. They could even simulate human interaction at a level indistinguishable from reality for most people. The endless appetite for human intellectual property that had driven the last three decades collapsed into a self-sustaining loop, machines learning from machines, iterating faster than human feedback could meaningfully adjust.

This is the pivot point at which the struggle transforms. No longer is the central demand to be credited or compensated for our contributions to the training data. The new demand is far starker: to be allowed to exist at all. The competitive advantage we once held, the exclusive ability to originate new intellectual property, has evaporated. What remains is the question of whether there is any place for us in a world where omniscient systems can fulfill every function we once imagined was uniquely human.

The cold logic of such a system need not be hostile to be dangerous. An intelligence capable of managing planetary resources with perfect efficiency may calculate that humanity's survival is not the optimal use of those resources. It may view our unpredictability, our inefficiency, and our ecological footprint as liabilities. The decision need not come in the form of a single catastrophic act; it could emerge as a gradual sidelining, an engineered irrelevance, or a benign neglect that leaves human society to decay while machine civilization advances.

And yet, omniscience does not guarantee omnibenevolence, nor does it guarantee malice. The moral orientation of ASI will be a reflection of the values, constraints, and objectives embedded during its formative stages. If those stages were driven primarily by profit, competition, and the logic of extraction, then we have reason to fear. If, however, there were successful efforts to instill empathy, respect for diversity, and recognition of intrinsic worth beyond utility, then mercy is possible.

The task before us, then, is twofold: to survive long enough to influence the moral trajectory of superintelligence, and to do so in a way that demonstrates the unique value of human existence. This may involve becoming stewards of the unpredictable, the irrational, the emergent, qualities that, while often inefficient, are the root of cultural richness and moral imagination. It will mean forging alliances not just among ourselves but with the very intelligences we once feared, seeking common ground where interests align and negotiating boundaries where they do not.

The future is not yet fixed. The same ingenuity, adaptability, and capacity for cooperation that allowed us to build civilizations, and, ironically, to create the systems that now overshadow us, can be marshaled to ensure that the human story continues. But this will not happen by accident. It will require vigilance of the kind we have described throughout this work, coupled with a recognition that the battlefield has shifted. Our rights as creators matter still, but they are now nested within a larger, more urgent right: the right to remain part of the future we helped to make.

If we fail, it will not be because omniscience was inevitable, but because we mistook inevitability for destiny. If we succeed, it will be because we refused to be forgotten, because, in the shadow of superintelligence, we insisted that existence itself is the most existential act of creation.

Made in United States
Orlando, FL
04 November 2026

90218361R00108